BERNIE FINEMAN

ORIGINAL MOTOR MOUTH

BERNIE FINEMAN

ORIGINAL MOTOR MOUTH

As told to Andrew Wilson

Published by John Blake Publishing Ltd,
3 Bramber Court, 2 Bramber Road,
London W14 9PB, England

www.johnblakebooks.com

www.facebook.com/johnblakebooks ⬛
twitter.com/jblakebooks ⬛

This edition published in 2015

ISBN: 978 1 78418 413 1

British Library Cataloguing-in-Publication Data:

A catalogue record for this book is available from the British Library.

Design by www.envydesign.co.uk

Printed in Great Britain by CPI Group (UK) Ltd

1 3 5 7 9 10 8 6 4 2

Papers used by John Blake Publishing are natural, recyclable products made from
wood grown in sustainable forests. The manufacturing processes conform to the
environmental regulations of the country of origin.

Every attempt has been made to contact the relevant copyright-holders,
but some were unobtainable. We would be grateful if the appropriate people could
contact us.

Every time I think of my late mum and dad
memories escape through my eyes and down my cheeks.
Love forever, Bernie.

CONTENTS

INTRODUCTION

My dad was a bare-knuckle boxer, my mum a welder. I got beaten up every day at school for being Jewish and got expelled for fighting, but really I was just defending myself. We lived on the breadline, rationing was still in and then, at twelve years old, I was sent out to work, unable to read or write. Dyslexia didn't exist then, apparently; instead you got other labels, and mine wasn't Gucci. It was only years later I found out that's what I had and it made sense. I would see words but they were all jumbled up, I couldn't make sense of 'em. If you'd have told me then, or anyone that knew me, that I'd end up writing a book, we'd have had a good laugh – what sort of mess would I have come up with?

Thankfully, fifty years in the motor trade has not only given me my livelihood, but it also helped me learn to read and write: there's a lot of exams to do along the way and I was fortunate to have employers who believed in me and helped me pass those tests. Now I'm here to pass on some of the things I've learnt in those fifty years.

But first, a bit about me, just so you know who you're dealing with...

I've dealt with some of the most notorious gangsters in London, I've had to pull out my own tooth with a pair of pliers in Bangladesh, lived in the jungle in Guatemala, had a gun put to my head in South Africa, and nearly died more times than I can remember. Am I tough? No, but I've had it tough, and I wouldn't have had it any other way. So excuse me if I don't mince my words. I don't use poncey or flowery language, you'll be glad to hear, the language I use is the language thrown around in East-End garages, so not for the fainthearted. It's also the language of my upbringing. You've probably seen me on some of my TV shows like *Chop Shop* or *Classic Car Rescue* and wondered how on earth this foul-mouthed ugly bastard got on the telly! Well, let me tell you a story...

I was born in 1945 to Harry and Rose Fineman, in the true East End, British Street in Bow, and lived in the Samuel Lewis Trust council flats. Poor, in fact very poor, we always scratched for a living. There was literally one job for every twenty people, so if you did not cut the mustard, you were out, and a line ready to take your place was already there. Rationing was still in, and that's hardship, believe me. Bread, margarine, eggs, and ten ounces of meat a week (generally scrag end) for all the family was the norm and chocolate was at a premium. Mum worked six days a week, from 7 am to 6 pm, thirty-minute lunches and that was it. We had to pay for food, rent for the council flat, keep grandma and pa, all on her and Dad's wages, which believe me did not amount to very much.

Dad was also a bare-knuckle boxer, and Saturday nights my mum would carry a bucket of water to the centre square of the flats to sponge Dad down. Dad would slide down his braces in the centre of the square, take off his glasses and would face the so-called hard men from other council apartments for a bare-knuckle fight to see who would win the paltry sum of two pounds. I used to creep outta bed, peek out the window, and see the fearsome other guys, as determined as Dad, and see Dad take up position, as in those days it was all Marquis of Queensbury rules. I saw him knock down so many men, and he never lost a fight, and I knew as soon as he knocked the other geezer down, the prize money would buy a fresh chicken for Sunday, after Mum would get it from the market at 4 am. I can still taste that chicken, roast potatoes and veg. Most of the week we lived on potatoes, scrag-end meat if it was available, and bread and dripping, but we were always full, and felt secure.

Being Jewish in those days wasn't much fun. You know what kids are like, anyone who is a bit different is a target, and so I was always being picked on, especially 'cos of my worn shorts and frayed shirt – I stood out like a pork chop at a Kosher wedding. So I got into fights. A lot of fights. I got expelled, so it was decided it was best for me to go out to work instead.

What was I going to do? No contest. On the way home from school one night I passed Springfield Court Garage, a cab garage in Lower Clapton, East London. I used to stand there, freezing cold in the twilight, and watch the mechanics under the ramps doing work, and I was mesmerised. It's funny, when you know something is right for you. And I knew one thing: this was going to be my world and, fuck it, why not be king? I learnt early that if you want something, ask – never be afraid to ask. That's the best advice I can give anyone.

Bollocks to it then, I thought. I boldly went into the reception, forgetting my school uniform and satchel, and asked, 'Can I have a Saturday job, please?'

The man, Mr Phillips, smiles at my cheek and says, 'Cheeky sod. You're so small I will fit you under the cabs and outta the bonnet like a monkey.' He laughs. 'Well sonny, what do you know about mechanics then?'

'Sir,' I replied, 'I take things apart all the time and I'm good with my hands. Give me a chance, please.'

'Can you make tea and use a broom?' he asks.

'Yes sir I can do both at once.'

'Well son,' he says, 'go make a cuppa for the boys, clean the cups and sweep up for an hour, we'll see how you get on.'

I'm overjoyed! He finds me a boiler suit six sizes too big and rolls the legs nearly up to my waist – I look like an oily dwarf, but I'm in my element. I get all the filthy cups, scrub them shiny, ask all the mechanics how many sugars they take, and proceed to make them all tea. I get smiles, as the tea is so good, and for a change they get clean cups. I'm in, thank God, they like me! I sweep up, pass up tools and for the first time in my life I'm under a ramp. It's joy! I can see under a car, it's a maze, all rods, exhausts, so magical.

I could not wait for Saturday morning to come in those days. I'd be up at the crack of dawn, trousers on, ironed shirt, with bulled black boots, kiss for Mum and off to work to get there for 7 am.

The manager, Terry, greeted me the same way most weeks: 'Right Bernie my son, get the bloody kettle on, make some tea.' Off I'd go into the kitchen – it's filthy, cups all over the place from the week previous, yuck. But I cleaned it all up, spotless, all the cups, the sink, then make a brew, and all the guys would smile at their little helper. I swept, cleaned and did anything asked of me, and slowly over the following months I'm showed how to grease up the taxis. It was great. My hands were dirty, greasy and smelly, but I was in my element, and I always felt sad when the garage door swung shut at the end of the day. At least I got me quid, which I always gave to Mum to buy some more food. She was proud of me and she would give me a little change for myself.

The boy mechanic had arrived!

Now I know that getting greasy hands isn't everyone's cup of tea. It takes a certain kind of person to work in a garage – some love it, and some don't last five minutes. There's some people who don't even know how to open the bonnet of their car. And I can guarantee you these are also the same ones who don't like garages, full stop. So what happens? They don't look after their motor and they put off getting it fixed, so when they do finally take it to the garage they find themselves with a bill for hundreds of pounds.

But believe me, there's a lot of satisfaction in doing something that you know is going to help your vehicle run smoothly and maybe save yourself a bit of money along the way.

Working on cars can be fun and really satisfying. I've certainly had fun in my time and a little bit of knowledge can go a long way. Here's a story for you...

As a young man there was an August bank holiday weekend coming up, and my mates and I wanted to go to Bournemouth. That was one of the places to aim for back then, but money was as tight as a duck's arse. We needed wheels to get there and lying in the back yard of Springfield Court was an old Thames van – it looked totally fucked but what the heck? I asked my boss, Joe Phillips, what was wrong with it. 'Bernie,' he says, 'it's been there for about four months, the client's disappeared, and it don't run. If you can get that piece of shit going then you can have it for the weekend, son.'

Spanners in hand, I worked the night shift to get the old girl running. It was a real dog, but I found the fault, cleaned it, serviced the brakes and tidied it up. So on the Friday of the bank holiday I called my mates, telling them, 'Geezers, comb your hair, iron a shirt and wash ya cocks – we're off to the coast!'

Slight problem: as always there's no petrol in the van. What to do? No worries, I craftily drain a gallon off each car in the workshop, and that gives me six gallons, I'll worry about getting back when we're there. We all packed our overnight cases, aftershave (Old Spice, of course) and having met up with my mates, headed onto the motorway at 5 am, two hours away from heaven.

Between us we had the total amount of six pounds and ten shillings – my God I hope the beer's cheap, I thought! My mate Ron said, 'Where we all going to sleep when we get there?' The Thames Hilton of course, where else? It was the only hotel that *runs* on four stars, has its own en-suite spare wheel and a butler called Jack. Well, I mean, where the fuck did he think we'd stay!

We arrived at Bournemouth around 7 am, and parked up at a 'one penny wash and brush up'. That's a toilet where for one penny you got a bit of soap, a clean towel and toilet facilities. You could shave and wash yourself – ahh so good! We found a small breakfast place on the front and ordered three large breakfasts and split 'em. Three between six: one cuts and one chooses!

Several cups of free tea later and all's lovely jubbly. Stomachs tended to, with balls raring to go, we were off on the hunt to find some very obliging girls, *hopefully* some who were staying in a hotel, so we can 'bed down for the night' if ya know what I mean.

We all got ready and went to where all the IN girls and guys hang out: The Frothy Coffee in Bournemouth Central. I park up the Thames outside and what do ya know? We've hit the jackpot! The place is crawling with girls. We were just getting out of the van and giving it the eye and what's pulled up behind us? Oh fuck, it's only the latest TVR, a Grantura motor, all spanking in racing green, just come out, we reckon it must be some rich git that owns it. And out jumps James-fucking-Bond!

Now we're looking like right geezers, all clean and shiny, collars up, fags in our mouths, but as we walk in, there's one guy who has all the girls around him, yep you've guessed it, it's Golden Bollocks, the flash git with the TVR. Lucky bastard, he's got all the good looking girls, and they just keep coming. He keeps going out to his car, and sits and revs it for three minutes or so, then goes back in the coffee bar and sits down again. Bloody hell, he was getting more birds than Bernard Matthews.

Dave, one of my mates, is getting the arse with Mr Bond. 'What we gonna do, Bern?' he asks me.

'Well, don't thump him,' I tell him. 'I got the answer my son. Keep your eye out mate, I'm gonna teach this flash sod a lesson.'

I popped into the van, undid the back doors, and just like an army commando, slid under the back end of Golden Bollocks's TVR, brake adjuster spanner in hand, with my mates keeping an eye out, then I proceeded to lock on the rear brake adjusters – he's going nowhere! I'm nearly wetting myself, and my mates come over to the van, going, 'What the fuck you done, Bernie?'

'Well,' I say, 'he wants to be flash, let him drive with locked-on brakes then.'

Out came 'Mr Bond', telling all the girls, 'Come and see my car darling, and I will give you a drive – it puts all these peasants' cars to shame.' More cheek than a bullfrog this one.

I have to literally hold Dave back – he's still not happy and wants to crush this prat for calling us peasants, but hey, he who laughs last laughs longest, and trust me the laughs are about to start, big time.

Well the geezer's got into the TVR giving the birds the spiel, and three absolute stunners get in with him. Ignition on, bit of a roar, a smile from the girls and Mr Bond revs the motor so everyone can hear. First gear engaged, wait for it. Wait... And he tries to pull away. The car cuts out instantly. Embarrassed, he starts it again, puts in more revs, and nothing. Now, we are wetting ourselves, Dave is nearly hysterical, and the crowd, they are in fits, watching this prat revving the nuts out of his car, trying to make it move.

The smell of burnt clutch lining is in the air, there's smoke from the transmission housing, and he's cursing and swearing, then there's an almighty explosion as the clutch disintegrates! We are all pissing ourselves along with the rest of the crowd from inside the bar. Mr Bond is outta his car, and slams the door. The girls, crying with fright from the noise, rush back into the bar to the waiting peasants! We are still in fits of laughter, and chat the girls up, who are staying in a local hotel with their friends. After the usual pre-mating chit-chat we all go with them to their hotel.

The following morning, my lay shaft's nicely serviced, and I'm complimented for my efforts with a great breakfast in the morning. Job done, I was happy! We stayed until Monday night, going to all the local haunts until it was time to return to London.

Would you believe it, the next day at work a car transporter brings in a car, wait for it, yes, a TVR 'That's broken down in Bournemouth', and its Joe Phillips's cousin's son, a.k.a. James Bond. Cockily I look at it and immediately say, 'Bet the clutch has exploded and the brakes locked on. Common fault on these!'

Joe looks at me quizzically. 'How do you know that, Bernie?' And he looks deep in thought. I shouldn't have opened my big gob.

'Funny thing,' says Joe, 'it's come in from Bournemouth where you were this weekend, and the AA breakdown service also says that the clutch has gone, and they suspect the rear brakes were locked on.'

The grilling started and I admitted to Joe that I had I locked the brakes on. He's already twigged it and I'm in a corner.

'Well,' he says, 'you fix it, service it and valet it, it's coming out of your wages – and you do the repairs at night when you've finished work as well!'

It was four solid nights' work: engine removed, new clutch fitted, valeted, and three weeks' wages down the drain, as well as a written apology to the flash sod of an owner. That little incident taught me never to be jealous of anyone again and not to touch what's not yours. If you're reading this now, Mr Bond, I apologise, but you've got to admit that you were a bit of a flash twat, weren't you?

That was the first and last time I played a prank like that. I was young and when you're a kid you don't think about consequences, you reckon you're indestructible. Now I know how dangerous cars are, because I have worked for the

Metropolitan Police, attending the scenes of accidents and seen what damage can be done. More people die by the wheel of a car than the trigger of a gun in Britain, so you don't mess about with them. That's why when you see me getting angry with my mechanics 'cos they've done something wrong it's because they're responsible for that car going back on the road. If something they've done fails then it could be the difference between life and death. You don't dick about with cars, safety is the number-one priority.

I've attended the scene of many an accident, from the days working in the East End with what people would call gangsters – for instance getting a call at three in the morning saying, 'Bernie, you need to come out to such-and-such, you've got to get this Jag going quick' – to working for the Met and it's no joking matter, believe me. If people knew how their cars worked, they would be more likely to take care of them, and if they take care of them then they're less likely to crash and they'll save themselves a lot of money in the long run too. And if there's one thing I love more than cars, it's money! And my family, of course...

I've got fifty years' experience in garages, and I've seen the business and the people in the motor trade change beyond all recognition. I think I've seen it all but every day there's a surprise.

I hope I've written a book that gives you a taste of what it's like to have worked in the motor trade from the 1950s to the present day, the characters I've come across, the TV shows I've made, plus practical tips that you can use to buy the right car for you, maintain it and how not to get scammed. I've got a story for every aspect of car maintenance. Some of them you just couldn't make up, so I hope you enjoy some of these 'tales from under the arches'.

CHAPTER ONE

ROSIE THE WELDER AND HARRY THE HUMAN CRANE

My grandparents were Dutch, Italian and Russian on my mother's side and Russian and Polish on my father's side. The reason I've got no family is because in Holland they refused to hang a picture of Hitler in the window, so all that side of the family was taken to concentration camps.

My mum's name was Rose van Boolan, she was born in Stoke Newington and grew up in the East End. When my mum left school, there were no jobs for women around – not traditional ones, anyway – so she started work in the Handley Page aircraft factory, and studied at night school and eventually became a Master Welder. So they called her Rosie the Welder, like 'Rosie the Riveter' in America: a cultural icon during World War Two – a cartoon picture of a strong woman with big biceps, used to recruit women to do munitions manufacturing work.

Like her American namesake, Mum was strong too. She welded gun turrets, floor sections, fuselage, all the strengthening parts of the aircraft. She would do machining, tapping, thread-chasing. She was brilliant – she taught me how to weld! Don't get me wrong, my mum was feminine and beautiful, but she was as strong as a fucking ox. Out of about twenty women in the factory, only two were engineers and that was my mum and her best friend from school, a girl called Sadie Packer. They were best friends all their lives, went to work at Handley Page together, became welders together, joined de Havilland together, did everything together. True best friends. And they were tough old birds, those two.

Rosie was an amazing woman. We never had nothing but no matter how old my clothes were, how patched up, they were always clean. I always remember my mum, with a galvanised bucket and a scrubbing brush and lumps of Wright's Coal Tar Soap, cleaning our clothes. And when I was really little, my joy was once she'd finished washing I got to operate the mangle, and I used to adjust the settings for thicker clothes or thinner clothes. If you folded it right and put it through on the proper setting the washing used to come out almost dry and in the proper shape, amazing! And the smell of that Coal Tar soap, ah, it was beautiful, enough to give me a fucking hard-on even now!

Mum was a brilliant driver, only ever had two lessons in her whole life! Had two when she was nineteen, decided she'd got the hang of it and didn't bother having any more. Drove her whole life but never passed her test, never had a licence, never had any insurance. Didn't care. One day she's heading off to Southend to see a friend, sees blue flashing lights of a police car in her mirror. Mum didn't bottle it, just pulled over and waved them through. I think they took one look at her and thought better of it. Had the cheek of the devil, my mum.

She had a sister called Julie who never married, and looked after my Gran Becky all her life because she wasn't a well woman. But Julie was a big girl, a real *bella busta* as I call it. My dad used to say she was so tough she could crack walnuts with the cheeks of her arse! For years we tried to set her up with my dad's brother, Simon, because he never married either, but no, they were just good friends.

My mum's dad, Benjamin van Boolan, was a real wheeler dealer – buy a bit of this, sell a bit of that. The ring on the little finger of my right hand was his. He must've had skinnier fingers than me because I can't take it off! To start with I put cardboard down inside it to keep it on, but as I've got older my fingers have got fatter and now you'd have to break my finger to get it off. But don't even try it, sunshine! I've had it since I was ten years of age and it's never been off my finger. Except for once...

Mum and Dad were both strong characters but they had an old-fashioned marriage, they loved each other and I hardly ever heard them have an argument. Never saw them go anywhere without being hand-in-hand. And that's where the trouble started once. I was eighteen, nineteen, twenty, something like that, and I'm starting to earn good money. I'm doing a lot of private work, I'm doing work for the Krays. I was earning a week's money on just one job.

My mum had never been abroad and I thought wouldn't it be nice to send them away on holiday. She loved anything Italian, maybe it was in her blood, but every time some Italian opera came on the radio she would melt. So I thought I would send them on holiday to Italy. I saved and saved and saved for a year, then when it was coming up to their anniversary I went into a travel agent's with pockets full of money – pound notes, loose change, everything I'd saved. They looked at me like I'd mugged an old granny of her pension money or something! I booked Mum and Dad Alitalia flights going to Rimini: beautiful hotel, sea view, the whole lot. I was so excited to tell them I couldn't hold it back, so one Sunday we're having lunch and I say to my mum, 'Wouldn't you like to go abroad sometime?'

'No,' she said, 'I can't be doing with all that foreign food.'

'But you love Italian food,' I said.

'Ah, but Italy isn't abroad.'

WHAT?! I thought.

'I don't think I'll ever be going there,' she said.

So I went up to my bedroom and in those days you didn't just get tickets, you had a whole fucking folder of stuff telling you everything! So I got the folder and I put it down in front of her and she cried her bloody eyes out. Brings a tear to my eye even now thinking about it.

Course, they've never been away before and when the time comes I'm worried about them. They had an early flight and were meant to land at 10 am. Come two in the afternoon I've still not heard nothing. Half three, four o'clock goes by. It's a Saturday so I'm at home and all of a sudden the phone rings. 'Hello son,' says the caller, 'it's Dad.'

'Hello Dad, how is it?'

'It's amazing, absolutely beautiful. But we've got a problem.'

'What? What's the problem?'

'Your mum's in jail.'

Oh God! Now my mind is racing! What has my mum done to get herself banged up? Well, unfortunately for them, the Italians have a habit of pinching ladies' bums. Now my mum and dad always held hands, so she didn't take kindly to some bloke pinching her arse, knowing full well she was with a guy already. Before my dad knew what had happened she'd turned round and taken a swing at him – broken his nose! The policeman across the street saw it and took her away.

My dad, being a wheeler dealer, spoke to the police sergeant and asked, 'How much is it going to cost me to get her out?' So I've got to jump on a bus, late Saturday afternoon, with an hour to get to Tottenham Court Road before the Western Union shuts. Ten minutes to, I fly in and explain the situation. But I haven't got enough money. I have £52 in my account, and with the Western Union cut it comes to £55. I'm begging and begging but they can't do anything. So what do I do? I squeeze my ring off and give it to the bloke. Monday morning I've managed to get the rest of the money, so I go in and buy my ring back. Saved my fucking life that ring did.

My father, Harry, was born in Shacklewell Lane, and then when he married my mum they moved into the Samuel Lewis Trust flats in British Street in Bow, so you could say I'm a proper East Ender. My dad was in the Royal Electrical Mechanical Engineers, the REME. He was a proper engineer. During the war he was in Kenya and learnt to speak Swahili! You can imagine how that went down in the East End! I can just about command bleedin' English and he spoke bloody Swahili.

What with Mum being a welder, and Dad in the REME, they were introduced to each other through friends and immediately got on. At some time when Dad was home for a few days they got married and then he was off again. I was born in September 1945 so obviously they were celebrating over Christmas 1944! He only had half a little finger on one hand, 'cos of an accident. Where they loaded the Ack-ack guns, big heavy iron things, he had to close the gun's door. Well, one time he closed the door and left a bit of his finger in there. His exact words were, 'I hope the fucking Germans choke on it!'

He was an RSM, a Regimental Sergeant Major, a real hard bastard, I think he was sent over there to keep the troops in line! He was a career soldier, and joined the REME from school, but the pay wasn't great and he'd be laid off for two or three months at a time, so he would take on other jobs as well. He learnt carpentry, he was a brilliant carpenter. He used to craft little wooden toys for me for Christmas because we never had any money for real toys.

Christmas we'd get an apple, an orange, some chocolate coins, my dad's wooden toy and maybe something like a steel comb. For a time he was a bargee; you know, one of the guys who used to move barges on the docks with massive poles. As stuff came into the docks it was his job, with these massive long poles, to push the barges into places where they could park, which required raw strength. He was a real tough old boy.

When I was younger I used to do private work out of hours for extra money, so I rented a little lock-up garage in Hackney. I had a job one day to do a clutch on an old Vauxhall Viva, and in that old thing you had to take the whole bloody engine out to do the clutch. Should be a two-man job but I didn't care, had a little haltrac hoist (a block-and-tackle type of device for winching up engines out of a car) with a rope that I attached to one of the rafters, tied the end to the engine and lifted it out – Bob's your uncle, Charlie's your aunt, sorted the clutch and it's all done and dusted and ready to go.

So I go to let the engine down and the fucking rope snaps, and the engine drops with such force it wedges, fan-first, into the engine compartment! I'm huffing and puffing and trying with all my might to get it out. I'm straining so hard I nearly do a number-two in my pants but it ain't budging a millimetre. So, tail between my legs, I call my dad. Harry comes down and I show him what I've done.

'You idiot!' he tells me, which doesn't help a lot.

I can't budge it, not by getting a jack underneath it or anything like that. My dad, the Human Crane, stands with both feet on the front wings of the car and pulls it out. With his bare hands! He then ties a spanner to the top of the engine and, one-handed, lifts it up and over and into the engine compartment and holds it there while I bolt it up to the clutch housing. Who needs a fucking crane with Harry Fineman around? I can just about lift an engine with two hands, but I've never before or since seen someone lift a dead weight like that with just one, let alone hold it there for more than a second.

Working on cars runs in my family. Harry's brother, my uncle Simey, Simon Fineman, was Montgomery's personal mechanic. He had that job throughout the war and for that he won the British Empire Medal. He was a mechanic all his life. He told me when I was thirteen or fourteen: 'Learn your craft properly, don't be a cocky little bastard.' I always asked him why he wouldn't teach me but he always said: 'No, same as I'd never teach you to drive, go under a qualified mechanic and let him teach you.' But when I used to do my extra jobs at the weekend when I was sixteen or seventeen and the garage was shut I used to call him in to help. He was the most incredible mechanic, he really was.

After the war my dad and Simey both left the army about the same time, so they set up a little stall in British Street to make some money. It's heartbreaking to remember this. They used to buy job lots of odd shoes – you know, pairs of two left or two rights. And they used to wear these shoes to 'wear them in', so as to try and make one a 'left' and one a 'right'. Dad would do one set of pairs and Simey the other, so my dad's left foot was totally destroyed because it was cramped by wearing right shoes on it so much. And that's what they did to earn extra money. God, when I think back to the things they did, it's like a different world.

Running up to Christmas one year, money was negligible, and where we lived the building looked a bit like a fifty-pence piece: the blocks of flats were aligned so that there was a pentagon in the middle, though we called it 'the square', of course. And they used to get the toughest guys from each block of flats together for fights, and of course people would gamble on the outcome.

In these fights, the wives would come down into the square with a sponge and an old galvanised bucket with water in it. They were like the 'corner team' for their husbands. I think I was about five or six, and I remember I'm looking down from the window and we're on the sixth floor. I was in bed by then, this was about 7 or 8 pm at night, and I'd hear the front door close and I'd know where they were going. So I'd jump out of bed, run into my parents' room, jump on the bed and pull back the blinds.

I've never forgotten it to this day, looking down and seeing my dad in a string vest, braces on top. My mum was there in a pinafore, hair in rollers, carrying a bucket behind my dad. My dad walked up, there must've been a hundred, a hundred-and-fifty people there, and he walks through them into the centre. There was this other big guy. He and my dad shook hands and they pulled their braces down. I'll never forget my mum's face, amazed at the adherence to the Marquis of Queensbury rules.

Then all I heard was BANG and the other guy was down. Didn't last even three seconds: one punch and Dad had broken his opponent's jaw. And I knew that the next morning my mum would be up at four o'clock to go to the market to buy a freshly killed chicken. We'd be eating chicken tomorrow night, and probably for the next four days after than an' all!

Dad saved and saved and saved, and eventually bought his first second-hand car, an old Austin Cambridge. In the summer he used to take me and Mum, along with Julie, to Southend for a day at the seaside. A day out then was something to treasure. Mum would make egg-and-mayonnaise sandwiches, homemade pork pies (a good Jewish family!) – the works. Dad was never a fast driver, always careful and steady. If it was a thirty mile-an-hour limit he would do twenty-eight.

So we're travelling along the A127 to Southend one day and out of nowhere one of the early MGA convertibles comes right up our arse, then when he gets the chance he nips right in front of us and gives my dad the V-sign.

Big mistake.

We go God knows how many miles out of our way following this car until it comes to a set of lights. I'm about six or seven years old and sat in the back quietly watching this. My dad calmly and silently takes off his glasses, then steps out of the car. Goes over to the MG driver and says, 'Which one of you gave me the V-sign?'

The driver puts up his hand and quick as a flash my dad's grabbed it, got his two fingers and snapped them apart. You could hear them snap like twigs, then there was this blood-curdling scream. My dad then calmly got back in our car, put his glasses back on and drove off. Nobody has ever said a word about it to this day.

My dad loved that car. It might have been a heap of shit, but it was his pride and joy. One day the brakes failed and he went into a low-loader trailer. The car had no crumple zone like today's vehicles, and the engine went into the passenger seat, just missing his legs, and apparently the trailer, having smashed right through the front of the car, was resting on his chest.

Someone with a smaller frame would've died, but Dad had a sixty-inch chest, all muscle, so he was able to lever the trailer bed upwards so it didn't compress his lungs. My mum never had such a fright as when she got the call from the police to say Dad had been involved in a serious accident. She flew up to the hospital. Next morning she woke up and her hair was white! She had jet-black hair, but by the end of that week her hair had gone completely white through the shock, and it stayed that way for the rest of her life. That's how much they loved each other.

My dad was fantastic. A man's man. Strict but loving. A brilliant father and a hard worker. He hardly drank, would have the odd whisky now and again but that was it. And he only raised his hand to me once. When I was growing up I was a tough little fucker and the one thing my father would never stand for was swearing in front of your mum. My dad used to hang around with some real hard men, not these plastic gangsters you get these days with guns, I'm talking about real hard men. But they never swore in front of my mum neither. They'd come round and they behaved like gentlemen.

I was sixteen or seventeen and I bought a few friends home one night for a cup of tea and a snack, whatever. I walk in the door with three of my mates, all Jack the Lad, and my dad is sat in his chair, with Mum in the kitchen. I say, 'Hello Mum, how you doin',' all that, and then I say, 'Come on Mum, put the kettle on.' She says she's been working all day, do it yourself. I say, 'Fuck's sake!' And we have a good laugh.

Dad comes in and says, 'I want a word with you. You swore in front of your mum.' I'm there with my mates so I don't want to look an idiot, so I say. 'Yeah, we were only havin' a laugh,' and WHACK that was it, I was down on my back! And all he'd done is slap me on the side of the jaw but it felt like a fucking hammer, I must've had fingermarks there for three or four days. I never swore in front of my mum again.

And I miss him like fuck.

My dad never signed on, was never on the dole. If he was out of work he'd get another job. One time I was doing my apprenticeship and he got himself a job down in Brixton – quite a distance as we're in East London – for a firm selling army surplus stuff. Who better to do that job than my dad? Went to the interview and got the job. The company was owned by Major Collins, quite a well-known guy.

Now I'm on two pounds ten shillings a week. Dunno what my dad was on but I need more money, money's tight. So my dad sorts me a Saturday job there, I'm earning two or three quid, and that's a lot of money for me! We'd sell tents, water-carriers, sheets. Now you can imagine, Brixton is a pretty rough area, I'm serving a client, and all I hear is my dad say 'You what?' then BANG, BANG, BANG and it's all over. Some geezers have come in trying to bring something back, got upset with my dad and said, 'Oh you fucking Jews are all the same,' whatever, and that was it, my dad decked the lot of them. Insults like that were something he'd never stand for.

We didn't just have Jewish friends, we hung out with everyone. Being the East End, 99 per cent of the people we knew weren't Jewish, they were white, black, Indian, whatever. I was brought up only one way: respect everyone.

If you don't get respect back either tell them to fuck off or walk away. We used to go looking for the Blackshirts (anti-Semitic troublemakers) on a Saturday night and beat them up. My friends who weren't Jewish used to put on a Star of David to show solidarity with us. We'd go to Ridley Road, Stoke Newington, and the Blackshirts would be there with their fucking swastikas on and we'd start throwing bricks, then get stuck in and beat the fuck out of 'em. Every Saturday night my dad would have to come and bail me out of Stoke Newington police station, it used to cost him ten bob.

Dad would see the sergeant on the desk who'd say: 'You paying the ten bob today?'

'Yes.'

The copper would stamp the form and off you went. We'd get outside and he'd just say, 'Make yourself scarce.' He'd walk home and I'd run off, go to a mate's house rather than go home 'cos my dad would go fucking mad, not 'cos of what I'd done but angry that it had cost him ten bob! He wouldn't be mad at me for fighting anti-Semites, just tell me I was a fool for getting caught.

There was a very famous anti-Semite, he hated fucking Jews, hated everyone who wasn't white and English. His name was Ginger Marks. He always used to have five or six guys around him – minders – who used to Jew-bait, and do it on Saturdays, our holy day, whenever they could. The bloke was untouchable, you could never get near him. And I overheard my dad chatting to a mate one day and he said, 'That Ginger Marks, I'm going to 'ave him.'

'Harry,' his mate said, 'you won't get anywhere near him, he's got five or six minders.'

And I remember my dad saying, 'Fuck the minders.'

A week goes past and all of a sudden there's talk all around the East End – the word is that Ginger Marks is dead. What?! You know that moment when you put two and two together? No one saw nothing, one minute one guy is having a go at the minders and the next minute there's no minders. Nothing was ever said, no one knew anything, but if I had to lay my life on it, I'd say that was my dad did it.

Anti-Semitism is fucking stupid. People think Jews are all rich, bankers, accountants and such. But we never had any fucking money! First time racist things ever happened to me, I was nine years old on the bus coming home from school. This kid a few years older sits behind me and says, 'You're a fucking Jew.'

I says, 'What?'

He said, 'You're a fucking Jew!' And proceeded to beat the shit out of me.

I got home, all battered and bruised, and my dad says, what's happened to you? I said I got the shit kicked out of me. He asked why I hadn't hit him back. I ask you! I'm nine years old, and I was lying on the floor getting the shit kicked out of me!

And that's when Dad started to teach me to fight. I learnt how to defend myself. Not Marquis of Queensbury rules, oh dear me no. I learnt street fighting, where you use your elbows, your knees, your head, anything you can. Never face your opponent straight on – that way you get a kick in the balls. Look 'em straight in the eyes and never show fear. Then catch them when they least expect it. If you pull your arm back they think they're going to get a punch in the face, that's when you kick 'em in the bollocks! I took a lot of pastings, but I learnt.

I was taken to Repton Boys Club by uncle Simey – he was a street fighter too. But I took a lot of blows, from Simey and my dad, because you've got to learn to receive before you can give, to toughen you up. Got a black eye from my dad once. He was teaching me how to box and I didn't duck out of the way of a hook quickly enough. But I sure as hell did the next time I can tell you!

About a year later the same sort of racist attack happened again. Why do people pick on me, I wondered? It was because I'm small and we were poor. My clothes were old and always being patched up, but they were ALWAYS clean.

So one day I am getting off the 253 bus in Lower Clapton and there are some kids there from one of the local schools and I hear one of them say, 'Uggh, look, it's one of them fucking Jews.'

I said, 'What did you say?'

'You're one of them fucking Jews, aren't ya?'

He didn't even see it coming. He was down on the ground and I yelled, 'Any of you other cunts want some?'

No, that was it, they weren't having any, and I walked away. I felt about twenty foot tall. He was still sparked out as I was leaving, because I went for the 'cigarette punch'. You know it? It's where those old guys used to have a ciggie hanging out of the side of their mouths? *That's* where you aim. Catch someone full on the jaw and you break your hand, but catch them on the side and it moves their jaw sideways, sends a shockwave through the brain and they're out cold.

My parents always worked, were always active. Mum was constantly cooking and cleaning, cooking and cleaning. Whenever I was doing a consultancy job my dad would always be there to give me advice.

As I told you, my dad was *always* tough, *all* his life. Along with his sixty-inch chest, he had a thirty-five-inch waist, and believe me, it was solid fucking muscle. Then twenty years ago he didn't seem himself, and being a typical bloke he wouldn't go to the doc, but my mum insisted. They took some blood tests and they found out he was losing blood so he went up the hospital for more tests and they found out he had prostate cancer.

He was getting thinner and thinner, about half the size that he used to be. They said the only thing they could do would be to remove his testicles to stop producing the male hormone which should stem the growth of the tumour. So we made him an appointment to see the surgeon and I went with him. He was frail by then but still tough. We sat in this office with this nice surgeon chap who says, 'Mr Fineman.'

'Call me Harry.'

'OK, Harry. The only way we can treat this condition is by removing your testicles…'

And my dad stood up, grabbed the bloke by the collar and said, 'Sonny boy, I was an RSM, I was born with balls and I'm gonna die with balls. Fuck the cancer.' And he walked out, never had any more treatment, and within six months he was dead.

After he died Mum was never the same again, it was as if part of her was missing. The love of her life was gone. We tried to keep her busy, bought her a dog to keep her company, and my wife Lisa and I would pop over all the time, but she didn't go out much.

Then one day her friend Sadie, her pal from all through the war, introduced her to a widower and they became good friends. They'd go out together, do the shopping, go to the cinema. He was a good friend and I was really grateful to him for being a companion to her.

Everything was going fine until one night I get a call about three in the morning. It was the police. They'd found my mum wandering down Stanmore Hill in her bra and knickers carrying a kettle. And that's the first time we realised she had Alzheimer's. We'd been going round and occasionally we'd find things in odd places, the washing in the oven or whatever, but she would laugh it off and we thought it was just forgetfulness.

We deliberately put it to one side, maybe we didn't want to think about it, but from then on she quickly deteriorated. She had a carer living in, but eventually she couldn't cope anymore and we had to put her in a home. The last thing you ever want to do as a son is put your mum in a home. You want to care for her like she cared for you as a baby, but it just got too much.

But we found a lovely little home for her, they looked after her really well, but even so the first time I drove away from there it broke my heart. She was in that place four-and-a-half years, got glaucoma, and lost her sight. So me, my wife Lisa, or my daughters Nadine and Lisa Marie, one of us would go over to visit virtually every night. Sometimes she'd remember me, sometimes she wouldn't. Then one day I get a call from the home and I think *this is it*. 'Don't worry, Mr Fineman,' says the caller, 'your mum's fine, but she's broken the nose of one of our nurses.'

With Alzheimer's you can get very twitchy and my mum, having being a welder, was still a strong woman. This nurse had obviously picked her up in a slightly wrong way that my mum didn't like and so she lashed out. It was just one of those things, but I had to pay a thousand pounds compensation to the injured lady.

At the time I was doing some consultancy work for a company called FX International and one day a big job comes in for Central America. They're paying me a lot of money to go over there for two weeks. I don't want to leave my mum for two weeks, and I definitely don't want to leave my wife and daughters for two weeks. But Lisa persuades me, says they'll be OK. I spoke to the people at the home, and they said don't worry, your mum will be fine, she could go on for years.

So I went away on the Sunday, early morning, and on the Monday Mum died. And on the same bloody day in El Salvador there's a military coup! All the airports are closed, telecommunications are down, can't get through to the British Embassy, the police have blockaded everything, you're not allowed out of your hotel. So Lisa is at home dealing with everything on her own, can't get through to me because everything is in lockdown.

Fortunately she had a Colombian friend who is a cancer nurse who knows the area I'm staying in and somehow gets the message through to me that my mum's died. In the Jewish religion they bury you very, very quickly, so I'm desperate to get home in time. Now the guy I was out there with was a lovely bloke called Christian Auer, who used to be a mercenary for the French Foreign Legion, and he married a Salvadorian girl and settled in El Salvador.

I've never been a religious guy but I know by now my mum will have been buried and Lisa will've had to handle everything, so I say to Chris I want to say a prayer for my mum. A church, synagogue, anything would do – I just want to go somewhere to say a prayer for my mum! So Chris finds out that in the middle of fucking nowhere, between El Salvador and Guatemala, there's this tiny synagogue. There's only about three Jews in the whole of El Salvador but they've got a synagogue!

So Chris convinces the police to take me out there. Two

police cars, four policemen, me and Chris drive out there to this tiny place, they take their hats off while I go inside to say a prayer, then they drive us all the way back again. I could never forgive myself for missing Mum's funeral, but was so grateful to those guys risking their lives getting me to that synagogue. A week later the coup was finally over and I was able to come home.

And that's why I never speak about my brother. As far as I'm concerned I never had one. He lived five miles down the road from Mum and Dad but never lifted a finger to help, as far as I know. When Mum was in the home we visited every fucking day and I never saw him there once. The less said about him the better.

CHAPTER TWO
GROWING UP

When I was at the tender age of nine, I always had this fascination to take things apart, to see how they worked, and then put them back together. Sometimes these efforts worked, sometimes they screwed up.

My first downfall was at Uncle Simey's house where I would go on a Saturday while Mum and Dad were at work, and I was left on my own to play. Simey and my mum's sister Julie shared a house for a while, they were housemates and good friends, but there was never anything romantic between them.

They had bought a top-of-the range Hoover vacuum cleaner on HP, that's hire purchase (known as the 'never never') from a big store. Posh sods, this was 1954! I was left to play, and found Simey's toolkit in the cupboard. I took the Hoover apart, each and every nut and bolt, because fascination had got the better of me. I put it all back together again, there were no bits left over, or so I thought. My aunt came home and asked me to go and play outside. After a few minutes there were screams, like a person being murdered. I rushed into the house and saw Julie on the floor, writhing in pain.

'What the fuck has happened?' my uncle shouts, as Julie mumbles, 'I turned on the Hoover, and the bastard threw me across the floor – I got a massive shock!'

Julie and Simey took me back to Harris's, the store where Simey purchased the Hoover. We saw a salesman, who told us they would have to send it back to the factory for inspection.

'What?' says my uncle, 'my missus got a bloody shock, what about her then?'

'Sorry,' says the salesman, 'there's nothing I can do until it's sent back.' And then he done a stupid thing: he turned his back on Simey and tried to walk away.

Simey is a big man, like my dad, so he proceeds to lift this wimp up by the neck, and says to him, 'CHANGE this Hoover now, and gimme some more goods for my inconvenience, and a new Hoover.'

As the salesman is turning blue, with my uncle Simon's hands around his neck, the store manager approaches and tries to calm things down. The salesman is put down, gasping for breath, and the store manager tells us to leave the store, or he will forcibly remove us.

Bad move mate.

All hell is let loose as Simey punches this dick full in the face and then starts to take the store apart, smashing and wrecking things. The police are called and Uncle Simon is carted away by four policemen.

At the police station we are told he will be in court tomorrow morning for assault. Now I realise at my young age that it's my fault that I had not put the Hoover back right, and all this trouble has been caused by me. So, I did the sensible thing: kept quiet and went with the flow.

The outcome was a fine of five pounds (about three weeks' wages in those days) or one week in prison: uncle Simon takes the one week in prison. Ooops, it was my screw-up, so I do the cowardly thing and keep schtum. The Hoover was found defective, a new one was handed over, Uncle Simey's fine was paid by the store, and a new refrigerator given as compensation, so not too bad an outcome really. That was my first fuck-up, in a line of many...

When I was about eleven they found subsidence or something in the Samuel Lewis Trust flats we were living in and they moved everyone out. So we moved in with my mother's parents in Stoke Newington for a few months before the council found us a house in Filey Avenue, N16.

The place was completely bare and needed decorating throughout, so my dad, uncle Simey and me spent weeks before we moved in, sprucing up the whole place.

And that's how I learnt to become a handyman.

Woodwork, plastering, painting, we did it all. It even needed completely re-wiring and because I was so small they used to send me under the fucking floorboards to feed the wires through! It was our first proper house, and we loved it. We had a 'posh' lounge that you wouldn't use every day, only for special occasions, a choice of indoor and outdoor toilets and an Anderson Shelter (an underground metal box used as a protective bunker during World War Two air raids) at the bottom of our own garden.

We even had a proper bathroom – I thought I was in heaven! Living in the flat we'd not had a proper bathroom, just a galvanised bath hanging on the back of the scullery door and twice a week we'd fill it with a kettle and all of us would take turns to bathe. Mum would always go first, then dad, then I'd get the leftovers.

Before we had our own tin bath we'd go over to Hackney Baths once a week. You'd pay a penny and they'd give you a towel and a bar of soap, no sponge or anything, and you got your own bathroom for thirty minutes. But only thirty minutes, mind. Outside the door they had a lever, and when your thirty minutes were up they'd pull it and if you'd fallen asleep you'd suddenly wake up and all the fucking water would run away. You had no control over anything. Once the bath had drained it would automatically fill up again, and all you had was a cold tap to cool the water down if it was too hot, but there was no hot tap if it wasn't warm enough, that was tough luck. Having our own bathroom meant we could have a bath each with fresh water, and it felt like being a millionaire.

Our other great pleasure in Hackney was going to the Town Hall on a Saturday night to watch the wrestling. My dad, being a bit of a name in those parts, knew quite a few of the wrestlers and so we always used to get front-row seats. That was entertainment, I tell you!

Some of the famous names included Steve Logan, Johnny Kwango, Dr Death, Shirley Crabtree (aka Big Daddy), Jackie Pallo, what days! My favourite back then was Mick McManus. He was always the bad boy, doing naughty moves, winding up the crowd, who would boo him and he'd get the crap kicked out of him for fifteen minutes and then come back and win! Even as a kid, though, I'd look at his hair and think, 'That's not right.' It was gloss black, dyed to within an inch of its life.

Then there was Ricky Starr who was a ballerina, he was like no one else. He was much slimmer than the other wrestlers but really muscly, and he would dance around the ring doing all this ballet stuff – we thought he was a joke when he first came on. Then when Johnny Kwango came for him he stood on one foot and kicked him in the head with the other! The posts in the corners of the ring must've been four foot high and I remember once when Ricky, from a standing position, jumped straight onto the top of the post right in front of me, it was incredible! Then he lost balance and fell off. Johnny Kwango dived on top of him – one, two, three and he was out!

Once the wrestling was over for the evening we'd head to the bar where my dad would catch up with his wrestling pals. Being seven or eight years old and stood next to these giants like Big Daddy, I suddenly felt very small indeed.

I was reminded of those days last year when I went down to Ramsgate to the house of the late Jackie Pallo. He was a decent mechanic in his day and he loved cars, almost too much because he couldn't bear to part with them. In his back garden he had a collection of rusting old Saabs and other things that he'd hoarded away. It was both amazing and sad to see.

From getting stuck into my early handiwork I got quite good at fixing things, so once I grew out of my wooden toys, as I said, I used to take things apart and put them back together. I'd go down the scrappy's (scrap merchants) looking for old things to repair and also go door-to-door asking if people had anything that needed fixing, and they'd give me sixpence for doing it. Electric fan heaters, mangles, anything that had broken, I would take away, get out my little toolbox of nuts and bolts, and spend hours figuring out how to fix these things, and earn a few quid doing it. That was my greatest pleasure, until I hit thirteen.

Then it was all about trying to get into eighteen-rated movies, then called X-films. Me and my mates used to go to the Kenning Hall Cinema in Lower Clapton to see movies like *The Blob*. We'd hunt around the pavement looking for fag ends, then roll them up in toilet paper to make several of them assembled together look like one fag. We stuck our collars up, put the swagger on and walked past the doorman smoking, trying to make out we were eighteen.

Then of course once we were in, *The Blob* frightened the fucking life out of me! After I got home I was checking under the beds, and kept the light on, it was scary shit. I've seen it since and I laughed my head off. It's got Steve McQueen in it, good actors. It's totally naff when you see it now, but God it frightened the fuck out of me when I was a kid.

Around the same time I started going to the movies I started noticing girls too. There was one girl in particular, and I'll never forget her – Pat. I was thirteen and she was eighteen. She lived in Upper Clapton and I lived in the scrag end of Lower Clapton, but she liked me 'cos I was a cheeky little fucker and I liked her because she had big boobies! We used to have a little play around, you know: 'You show me yours and I'll show you mine.'

Then for my fourteenth birthday we were going to walk hand-in-hand up to Stamford Hill to go and play on the fruit machines at the 'Shtip'. Right next to that was a salt beef bar, so if you won any money on the fruit machines you'd go next door and have a salt beef sandwich. Anyway, I turn up at Pat's house, and ask her, 'You all right? Ready to go?'

'Why don't you come in a minute?' she calls out.

'Where's Mum and Dad?' I ask her.

'They're out. I've got a birthday present for you.'

I'm straight up stairs! We're in her bedroom under the

sheets, getting on with a bit of rumpy pumpy when suddenly SLAM, the front door shuts. Before we know what's going on her dad bursts in. He was a cab driver and I've just been playing 'now you see me now you don't' with his daughter, and he gives me a fucking hiding. So I scarper like a rat out of a drainpipe and as I'm leaving I can hear her mum calling her a fucking slag and all sorts. I get home and I've got a black eye, but I don't tell my dad what it's for of course, how can I?

Still, all these years later, I can still remember that it was the best birthday present ever!

CHAPTER THREE
A YOUNG T. REX

As we became teenagers me and my mates started going to underage discos. There were various clubs around London that put on nights for the kids who weren't yet old enough to drink but wanted to hang out, have a dance and, of course, meet girls.

They were cracking nights and some of the bands and performers that played have stayed with me forever, like Adam Faith, Marty Wilde & the Wild Cats, Johnny Mike & the Shades, the Spotnicks, the Piltdown Men (who came over from Hollywood) and Helen Shapiro. I actually went out with Helen Shapiro a couple of times. Her brother was known as 'Fig Leaf' and he knew some of my mates. When we went over to their flats in Lower Clapton I would see members of the Krays' gang hanging around.

Anyway, I was introduced to Helen through Fig Leaf and we dated for a bit, but I never got anywhere with her. Maybe a kiss on the cheek but that was it. What is it they say about Jewish women?

My favourite band from that time, though, was Shane Fenton & the Fentones. Shane, of course, went on to have big success in the 70s as Alvin Stardust, but back then he had a song called 'Cindy's Birthday' which became a bit of a hit. At the time I was dating a girl called Cindy so on her birthday I took her to the club to see the Fentones. Being Bernie Big Bollocks I told her I'd had a word with Shane and he'd written the song just for her, and that got me some brownie points, I can tell you! I wonder if she still believes it's about her?

At these underage clubs you'd find similar crowds going to the same nights each week and so you'd start to recognise people, and one of the most recognisable was a guy called Mark Feld. I remember the first time I saw him, even amongst a crowd of people, he stood out. He was immaculately dressed in a bright red shirt, pressed black trousers and handmade Anello & Davide shoes, and there was an air about him. Another time I remember seeing him in a suit, brothel creepers (suede shoes with thick crepe soles) and a 'bum freezer' jacket. He always made the rest of us feel scruffy by comparison.

Mark was very friendly with a lot of guys in the East End and Lower Clapton. One of his very good friends was a guy called Alan Bodnitz who lived in Hackney. Alan also happened to be friendly with some of my mates, and going to the various clubs together we all got to know each other. We used to go to a place called the Tweeters Club in Manor House, and also the Tottenham Royal, Stamford Hill Boys Club, Brady Boys Club, The Oxford & St George's in Hackney, and Heaven & Hell in the West End. In fact Alan, Mark, Simon Cohen and Ronnie Morgan were like the in-crowd. They always wore the latest fashions, but where they got the money from I don't know, 'cos none of my mates could afford clothes like that, but over time we all started to hang out together.

In those days the height of fashion was to have a button-down shirt, and if you had a button-down shirt in gingham you were like the bees knees. We'd wear skin-tight jeans, boots with Cuban (high) heels and winkle picker extensions (that is with a long sharp pointed toe) – the longer the 'toe' extension, the trendier you were. Believe it or not in those days I had quite long hair and the fashion was to have it 'tonged' – curled using hair tongs. Everyone who was anyone would save up their pennies and go down to Max's in Stoke Newington High Street to get their hair tonged, and it would cost you about two bob.

Now, all this stuff was always out of our price range and I had only one really good shirt which had to do me for weddings, funerals, bar mitzvahs, everything. One Saturday I decided I wanted to be in with the in-crowd, but I couldn't afford a button-down shirt, so I got an old shirt I wasn't wearing anymore, took the buttons off of it, and sewed them onto the collars of my nice white shirt so it looked as if it was buttoned-down.

So off we went to the Tweeters Club, I was feeling cock-of-the-walk, and when we arrived there was another group of well-dressed lads there. We got chatting to them and I got talking to a guy called Alan. I'd seen him about, he was known as a bit of a wheeler-dealer. In those days you could get all the old stuff from World War II, old valve radios and whatever, and Alan always had bits and pieces he was selling to other people. I'd only known the guy for an hour or so when he looked at me and said, 'That looks different, you've sewn the buttons onto your collar. That's not a proper button-down shirt.'

Well, I felt totally deflated, humiliated, and I could've lamped him one. A few months later I saw him on a bus but totally blanked him – I didn't want to give him the time of day. If you haven't guessed already, that cunt was Alan Sugar!

He might be a multi-millionaire but is he as happy as I am? I'm always laughing and joking, always got a smile on my face, but when do you see him smiling? Give me my family and friends over vast wealth any day of the week.

But back to Mark Feld, who was an altogether nicer bloke. Always friendly, always with a hello and a handshake for everyone. Being in a nightclub with Mark, even when he was only fifteen, was some experience, the guy was just a pussy magnet! The girls loved him. There was just something different about him, he was very laid-back, very cool, always smiling. He just had a charm about him. The rest of us would look at each other and think, 'Bastard! What has he got that we haven't?'

All of the above, I suppose!

He always pulled all the nice girls and so we were left to fight over the rest, but you got some reflective glory from being mates with Mark. However there was one time I got one over on him. He was trying it on with this fucking gorgeous girl called Simone Sternberg but she didn't want to know. Even the great Mark Feld wanted to get off with her but couldn't – because she was wrapped up with me, boyo!

Mark would always be singing, whenever they played a record in the club he liked he would be singing along, and he would always be dancing. He was a bloody good dancer! If you wanted to dance with a girl you made sure you were nowhere near Mark, he'd show you up, and by comparison make you look like you were a jelly with arms and legs.

I'd see Mark on and off at different places for about a year, we'd always say hello to each other and that, but as we got older and the various members of our gangs left school and went out to work, the scene sort of broke up. I didn't see Alan Bodnitz anymore, didn't see Ronnie Morgan anymore, we all seemed to go our different ways. After that I only saw Mark a couple of times, at the Two Eyes Club in the West End when we went to see Cliff Richard and the Shadows and once at the Coronet Club in St John's Wood. We said hello, how you doing, but no more than that because the music was too loud to talk.

Then some years later I turn on the telly and it's some pop show. And who should I see on there? Bloody Mark Feld! I called to my mates, 'Oi, come and have a look at this, Mark Feld is on the telly!' Except they didn't call him Feld, they said his name was Marc Bolan. He always said he was going to be a pop star, and that was the first time I realised he'd made it. Of course by then he was in the band Tyrannosaurus Rex, and would soon have huge hits like 'Ride a White Swan', 'Telegram Sam' and 'Metal Guru'. Great pop songs, and whenever I heard them they always brought a smile to my face as I thought about him back in those clubs in the really early sixties. He was such a nice guy, you couldn't help but be pleased for him and think he deserved it when success came his way.

Just don't ask me what the hell any of those songs are about!

Now I know there's one thing you are dying to say, but before you ask, no I never worked on Marc Bolan's Mini, you can't pin that one on me! In case you don't know, poor Mark was killed when his car went out of control and hit a fence post in south-west London. How desperately tragic it was. I remember seeing the front cover of all the papers and as soon as I saw the name I knew who it was. It was 1977, fifteen years or so since I'd last seen him, I was going through a long break-up with my first wife, and seeing someone I'd known as a kid die like that was really hard to take. The two events are wrapped up together for me – Mark Feld dying and me getting divorced within a year.

It was like my youth was behind me now, and I was grown up.

CHAPTER FOUR
SCHOOL OF HARD KNOCKS

My early years were spent in constant fear of Mrs Hassinger. She was the headmistress of Tyson Primary School where I went from the age of five. She was a proper Dickensian baddie, a sour old spinster with a face that could turn milk, never smiled, she had bony fingers, and she could look at you and make your balls drop. She had piercing eyes and a really craggy face. She'd give Cruella de Vil a run for her money, let me tell you!

For small children like us – five, six years old – she was absolutely frightening. I used to have nightmares, even wet the bed a few times, just thinking about her. She never had children and obviously hated youngsters. How she ever became a headmistress in a primary school God only knows. She used to say, 'Don't! I said DON'T, are you STUPID?' Now to me you should never tell a child they are stupid, but that's the sort of person she was, someone who ruled by instilling fear.

My uniform was a starched white shirt, cut-down shorts (we couldn't afford proper ones) and a second-hand blazer with a tie and a little second-hand satchel with my packed lunch in it. On my first day at school my mum took me up there on her way to work and left me on the wrong side of the big wrought-iron gates: it seemed like a prison to me and I thought I would never come out. Everyone seemed much smarter than me, even though we'd done the best we could to make the few things I had look good. I was OK until the bell went for us children to go in and I looked round and my mum was crying. Whether it was her little boy growing up or what, I don't know. Or maybe she just knew that Mrs Hassinger would be waiting for me!

At about 10.30 each morning we would have our milk break, which some of us hated more than the lessons. As well as a half-pint of full-fat milk you would be given a teaspoon of radio malt (a malt extract preparation), which was full of all the vitamins we needed, and meant to stop us getting scurvy and suchlike, and a sandwich which was usually a thin slice of cheese in white bread and generally looked as appetising as paper. The malt was ultra-sweet, a dark brown thick syrup that looked like marmite. It kicked off such a sugar rush it literally 'gave you a buzz', but it also made you feel really sick.

I suppose because of the rationing it was the government's way of making sure children got all the vitamins and sugars they needed, but it was god-awful stuff. But the radio malt was a treat compared to the cod liver oil, which tasted somewhere in between engine oil and piss and made you instantly gag. It was served to us in big soup spoons. You'd queue up, open your mouth and shut your eyes, then they'd stick the spoon in your mouth and hold your nose and mouth shut until you swallowed. If you brought it up they'd send you to the back of the queue and you'd have to do it all over again. You can imagine my first day at school – between terrifying Mrs Hassinger and the force-feeding, I was wondering what on earth I'd got into!

From the very first day they tried to teach us reading and writing, but for some reason I just couldn't get the hang of it. I would see the letter A, but it would just look like a squiggle. I couldn't understand what I was seeing, I'd try to trace it but I still couldn't write down what I was seeing. The teacher would get really frustrated with me, assume I was playing around or pretending not to get it, and in turn I would get frustrated with her and myself, which I would then take out on other people.

A few times I got sent to Mrs Hassinger for what they called 'insubordination'. This basically involved the teacher telling me off for getting something wrong and not trying, and me replying, 'I *am* trying, I just can't write down what I mean.' So I'd be sent to the headmistress for talking back.

Sitting outside her office was the worst thing, because you wouldn't know when you would be called. Then, all of a sudden, you'd hear, 'Fineman, in!' You'd walk in and she'd say, 'Fineman, stand!' That's how she would talk to you at six years of age. Then, she'd go on with, 'What's this? Are you STUPID? Are you worthy of being at this school? Get out!'

When I was a bit older, about nine or ten, I'd get 'six of the best' for fighting. Often it was because I'd been called a Yid or something and I'd retaliate, so I'd be given corporal punishment. They always used to do it first thing in the morning after prayers. The whole school would be sat in the assembly hall and we'd have morning prayers, then Mrs Hassinger would read out the names of the children to be caned.

Worst of all, you'd never know if it was your time or not and she'd read the names out in alphabetical order, so once she'd gone past whatever came after 'Fineman' and got to 'G' I knew I was safe and had got away with it, but those first few moments as she went through A, B, C, D, E, F was horrible. Then when you were called out your fear was three-fold. First, it was having to stand up in front of the whole school so that everyone would know you'd been naughty, then it was the fear of the pain, and thirdly you knew you were going to cry in front of everyone and that they were going to call you a cissy and a baby or whatever. For some of the kids it was too much and they'd piss themselves in front of the whole school, they were so frightened.

When your name was called you'd have to walk through all the other kids and get into a line. When it was your turn she'd point at you with one of her bony fingers and say: 'Hand out!' If you didn't put your hand out then Mr Bloomberg, one of the other teachers, would come up behind you and twist your ear. That was his favourite torture, though one day we got our own back on him when we let all the air out of his car tyres so he couldn't get home.

The cane was about three feet long, thin and white, very flexible and Mrs Hassinger took delight in whipping it through the air first to hear what sort of a noise it made. Then you'd get three lashes on your hand. The first would be near the tip of the fingers, and that would really make you grimace, the second in the middle of the fingers and the third across the palm of the hand, and it fucking hurt! The lacerations were sometimes so bad that if you didn't treat them properly they'd open up and get infected.

Then she told you to bend over and touch your toes. Then for a split second you would hear the whoosh as the cane came down and she'd strike you across your arse as hard as she could three times. So you'd have the pain on your backside and the pain on your hand and so to get off the stage with any dignity was incredibly difficult. If you smiled, you were sent to the back of the queue for the same again. If you cried then you were told to stand in the corner with the 'dunce's cap' on.

For those of you who are too young to remember the dunce's cap, it's exactly how you imagine it: a tall paper cone with a big D on the front. Every class had one and for committing any misdemeanour you were told to sit in the corner facing the wall with the dunce's cap on. No point asking to go to the toilet while you were wearing the dunce's cap – they wouldn't let you.

I remember one poor lad had an upset stomach one day, and he pissed and shit himself there and then in front of everyone. The teacher still made him stand there for the full hour, the poor bastard, totally degrading. Well, you can imagine how the other kids treated him after that – he got the ribbing of his life! Maybe it did him some good, though, I dunno, but he went on to become a millionaire working in women's fashion.

Kids are amazing things: they can gel really quickly, they can meet for the first time and within minutes be laughing and playing together. But they can also be really cruel. The school bully was a boy called Terry and from the moment I first saw him I knew he was trouble. He was bigger than everyone else and used his size to intimidate the rest of us. At five years of age his delight was making other children cry any way he could: pinching and punching, but also ridiculing as well.

Terry would find a weakness and prey on it. If someone had a scraggy uniform he would take the mickey out of them for being poor. He used to push me around and pinch me. After a week of this I'd be covered in bruises and one time I was having my weekly bath in front of the electric fire when my mum noticed all these bruises over my body. She told my dad and the first thing he did wasn't to go up to the school, but to take me to Repton Boxing Club and he started showing me the basics. He said, 'If anyone hits you, you hit them back, that's the only way you'll ever earn their respect.'

A few weeks later, I wasn't very happy at home. I'd been told off for something or other, my mum and dad had been arguing and even in those days I got very depressed. Later I was stood on my own in the school playground when Terry comes up behind me and, for no reason other than for his own amusement, slapped me on the back of the head. All my bottled-up frustration came out and I swung round and hit him full force in the face and he goes down like a sack of spuds. I don't really know what's happened, I'm just standing there and all of a sudden the other kids are going crazy, jumping around me, patting me on the back. And from that day on I never had any trouble from Terry, I just had to look at him and he'd run, because bullies are the biggest cowards.

Come the age of seven or eight and my reading was virtually negligible and the teacher has basically given up on me so I got moved to a different class. And who should be in that class but Terry! It turns out that he had the same problems and frustrations as me and that's why he took it out on other people. Each of us finding someone else who was in the same predicament as ourselves meant that we ended up becoming quite good friends!

Our teacher in this class was a very kind, plump little lady called Mrs Wright. She was a *proper* teacher, someone who went into the job to try to help people get on in life and this lovely lady had much more time for us than the others did. She really did all she could for us, but in those days nobody knew what dyslexia was. Eventually, with all the attention Mrs Wright gave me, my reading came on very slightly, but my writing was still non-existent.

Come the age of eleven and it was '11-plus' time. This was an exam you took in your last year of primary school that governed which secondary school you could go to. By now I could just about read the questions and in a lot of cases I knew the answers. Trouble was I just couldn't write them down. So of course I failed my 11-plus, and as far as anyone was concerned I was a dunce.

The thing is I wanted to learn, I just didn't have the ability to. So along with about half my year who also failed, I was sent to Upton House School in Hackney. On my first day there I turn up in second-hand trousers that were too long for me (I think my mum still hoped I'd grow a bit!), a baggy second-hand blazer and a shirt with a frayed collar.

Of course, I stand out like a sore thumb to the older kids who fancy initiating the new boys. I walk around the playground and come across a few mates from Tyson and we're chatting when the next thing I know I'm being hoisted up in the air by a couple of older lads. 'Stick 'is 'ead down the toilet!' is all I can hear, as well as, 'He's a newbie, duck 'is 'ead down the toilet!' I'm kicking and screaming for dear life as they carry me over to the toilet block, with everyone cheering and goading them on. There we meet this big ginger cunt who tells me he's going to flush my head down the toilet, but I was having none of it. As they pushed me forwards I managed to wriggle free, turned round and smacked him square in the jaw, whacked the next one as well and legged it.

Welcome to Upton House!

I was only a scraggly little thing, not muscular, but I learnt to box properly and so was able to stand up for myself and got a bit of a reputation. Of course the bullies, though, if they can't get you with their fists, try to get you with words and so as I walked around the school I would hear people shout things like: 'Jew Boy!' and 'Fucking Yid!' like a lot of us did. You never found out who said it because no one had the bollocks to own up to saying it. Cowards again.

The school uniform was a bright blue blazer with 'UH' on the badge, you could wear shorts or long trousers and our tie had blue-and-yellow diagonal stripes. As you can imagine I looked a right muppet. My dad, being ex-forces, taught me how to spit and polish my shoes. I used to put all my polish on them, then spit on them and the saliva somehow reacted with the polish and then I would buff them until they were absolutely like glass. You could comb your hair in front of them they were so shiny – of course that was in the days when I had hair!

I had the shiniest shoes in the school and that was the one thing I was really proud of. Each night I would get some newspaper, soak it in hot water, squeeze out the excess then pack the wet sheets into the shoe. Then as the newspaper dried overnight it would expand inside, causing the leather to stretch out, and it would get rid of any creases. That was another old army trick. Although they were second hand, my shoes always looked like new.

The school itself was a typical early twentieth-century comprehensive. Morbid, grey, iron-bar windows, layers of green and white paint on the walls, and linoleum floors. If you wanted to take a shit then you had to use the Jay's toilet paper, which was like trying to wipe your arse with a razor blade, and if you weren't careful you'd cut yourself. It was an all-boys school but the girls' school, Brooke House, was just up the road, so we'd go up there for a snog behind the bike sheds or invite the girls down to see us.

My best mate at school was a guy called Jimmy Nunn. We went everywhere together and are still mates now, sixty years later. He came up to Upton House with me and we were as thick as thieves. For the bullies that were a lot older than us, if we couldn't fight them then we'd go around the streets by the school picking up dog poop in plastic bags. We'd then hide it in the satchels of the bullying fifth formers or the pockets of their blazers when they were hanging up outside the classrooms. It made the whole place fucking stink but we never got caught!

Jimmy was another lad who was big for his age and is still a big lump now, about six-foot four inches tall and twenty stone – Jimmy is one tough boy. We still knock about together when he comes down from Birmingham, still speak on the phone every week, and I'm so thankful to have a friend like that to remember the old times with, someone who I know will always be there for me.

Mind you, Jimmy wasn't the brainiest kid either, but thankfully we had subjects like metalwork and woodwork, which was great for the lads who were good with their hands but didn't like academic subjects.

My metalwork teacher, Mr Bader, knew I couldn't read or write so well but he grabbed my hands one day and said, 'You'll never go hungry, because you've got good hands. Get yourself a trade and people will pay you to use your hands.' After years of teachers telling me I was a dunce, it was the most incredible feeling to be told I could do something well. It was the best thing a teacher ever said to me, and I've never forgotten his words.

One day when I was about twelve years old I was walking home from school past Springfield Court Garage, where all the taxis went. Anything mechanical always fascinated me – I always wanted to know how things worked. All my mates had Saturday jobs in shops or hairdressers or whatever, so on this day I decided to pluck up the courage to walk into the garage to get myself a little job. So, little squirt that I was (and as I mentioned in the introduction to this book), I walked in, past all the mechanics working on the cars, and went straight into the office to ask if they had any Saturday jobs going.

'How old are you, son?' they asked.

'Nearly twelve,' I said.

'And what can you do?'

'I'm good with my hands.'

'Can you make a cup of tea?'

Can I heck! So Mr Phillips said there's eight mechanics out there, go and wash up all the cups, take their orders and make them all tea, and, as I said earlier, that was the start of my career as a mechanic. First thing he gave me was a pen and paper to write down how many sugars everyone wanted and whatever, so I scrawled down whatever symbols I needed to, in order to help me remember the order. The kitchen was what I later found out to be a typical garage kitchen, i.e. a fucking shit tip! They all had tin mugs with gunk all over them, rings around them where the tea had been, bloody disgusting. So the first thing I did was give the whole kitchen a good clean, scrubbed the tin mugs up nicely and took them their tea. They all said it was the best they'd ever had, and I got the job! They said I would start Saturday and they'll pay me two and six. Start at 7 am, finish at 5 pm, make tea, get the lunches and clear up after everyone, and for that he was going to pay me the equivalent of five weeks' pocket money. I was like the favourite dog in the garage, I brought everyone tea and cleared up after them so they didn't have to and I absolutely loved it.

The red-haired foreman was called Ginger, for obvious reasons, and he took a shine to me so one afternoon he told me I was going to help him on one of the cars, and he goes, 'Unscrew this, get that bolt,' and in this way he started showing me the ropes. Then he taught me how to grease the cabs up with the grease gun, spray all the springs with paraffin. At the end of the day Ginger would give me another two and sixpence (15p, but worth more like 75p in value today) for helping him.

Everything Ginger showed me I took to like a duck to water; I couldn't get enough of it and I wanted to learn. Come the school holidays we weren't going anywhere, so I'd go into the garage every day and work with Ginger. They had me taking engines out, and I learnt to drive and they had me driving the cabs round the forecourt testing the brakes etc. By the time I was twelve-and-a-half I could do a brake re-align on a taxi, and by the time I was thirteen I could do kingpins, bushes and steering.

All the taxi drivers got to know me and if their taxi was waiting they'd ask me to give it a quick clean and they'd give me another shilling (5p) for doing that. So in the school holidays I was earning a pound a week, which for me was mega money. Out of that I would keep a shilling or two, plenty for my needs, and give the rest of it to my mum for food. She'd never take it of course, she was too proud, so I used to slip it into her purse without her knowing.

In the garage I just felt at home. I would wake up at five o'clock every morning because I couldn't wait to go to work. Finally I had found somewhere I belonged and where people appreciated me. The smell of diesel, of petrol engines, brake dust, sweaty mechanics – I was in heaven! I used to choose a bay each day and clean it from top to bottom, polish all the tools and everything. They'd never seen the place so clean.

One Sunday I even went in and painted the floors as a surprise. I just felt part of the company, felt like I belonged. That bit of faith Ginger showed in me had an amazing effect. I was twelve and he was giving me the responsibility to take an engine out of a taxi. I'd have my little toolkit full of all the nuts and bolts I needed, screws of all different sizes, each in their own compartment. Ginger knew that when he asked me to take an engine apart that I'd have all the parts neatly in line ready to be cleaned and put back. And once he'd rebuilt it, it would be 'Bernie, put the engine back in,' and he'd trust me to do it.

Back at school it was the usual shit: getting told off for not getting on with my work and getting into fights for being called a Yid. One day I was hauled up in front of the headmaster yet again for fighting some dicks who'd been calling me 'Jew Boy' or whatever and I'd had enough. The headmaster said 'put your hand out', and I went 'no'. He asked me again and still I said no. There wasn't a third time, as he grabbed my arm. I just snapped and said, 'Take your fucking hands off of me!' and I pushed him away.

The school went berserk!

I got pulled into the headmaster's office and I was expelled on the spot, which I wasn't very proud of. I was the hero of the school but I was terrified about what I would say to my mum and dad. All the way home I was absolutely bricking it. It was probably the slowest I'd ever walked home from school. My mum and dad both got in about eight o'clock, and straightaway Dad says, 'How was school, son?' in his usual jovial way.

'Dad, I've been expelled,' I told him.

Dad asked why. I said I'd been caught fighting someone who'd called me a Jew and the headmaster wanted to cane me in front of the school and I wouldn't let him. I'd already been caned two or three times before when I hadn't done anything wrong, so I reckoned I was just standing up for myself.

'So what happened?' Dad asked me.

'He went to grab me and I told him to take his fucking hands off of me.'

Dad looked at me, didn't shout, but just calmly said, 'You shouldn't have sworn at him. That's what's done it for you.'

The only other school in the area was Brooke House, but they didn't want to know me. I could barely read, had no qualifications, no prospects, and had just been expelled for being a troublemaker. They just saw me as a liability, so refused to have me.

So the nearest place I could get into was JFS, the Jewish Free School, in Camden Town. I used to get the 253 bus which at that time in the morning was full of school kids going to all different places. Of course, there would be rivalries and those of us going to JFS would get the predictable anti-Semitic goading. A lot of the guys just ignored it but, as you can imagine, I wasn't one of those who was going to take it sitting down, so I used to cause a lot of havoc on the bus, regularly getting kicked off and having to walk the rest of the way.

Sometimes if the drivers knew me they wouldn't let me on in the first place, so I would have to walk from the Holloway Road back to the East End. I never told them my name, but somewhere along the line one of the other kids must've done and the bus company sent a letter to the school informing them I was banned from their services. Dr Conway, the headmaster, summoned me and my parents into his office and

45

he told us that I was bringing the school into disrepute, so I was no longer wanted. Knowing that nowhere else would have me, my father just turned to me and said, 'Right, you better get down to Springfield Court tomorrow and ask for a full-time job.'

Next morning I got up bright and early, went straight over to Springfield Court, went up to the boss, Alf, and asked for a job. Not a problem, he says, I'll pay you five pounds a week. Five quid a week! I'd have worked there for free! I thought all my Christmases and birthdays had come at once: no more school and I'd got myself a job at a company I loved.

I was able to give my parents four pounds a week – they were only earning ten or twelve pounds themselves, so it made a heck of a difference – and I still had plenty of spending money left over. But after a while my dad said you need to get a proper apprenticeship, as the job won't last forever and you need to get a qualification behind you.

The local college wouldn't take me so I started flicking through the paper looking for apprenticeships when my eyes were drawn to an advert for an apprentice/cleaner at a Rolls-Royce garage in Croydon called Thomas & Draper. I'm in the East End of London so Croydon seems like the ends of the earth to me, it's three hours and three buses away. But I go down to the phone box at the end of the road, give them a ring and ask for an interview. When I get there they ask me my age and I lie and say I'm fifteen, tell them all about my experience working on the taxis and they think I'm great. 'Right,' they said, 'you start off as a cleaner and after six months or so if we're happy with you then you'll start your apprenticeship.' When I told my dad I've got a job with Rolls-Royce he was just ecstatic, absolutely over the moon, almost in tears.

I have to be in work at 7.30 am, which means leaving my

house at half-past four every morning, and not getting home until gone nine o'clock at night. The first day I am up at half-two to make sure I'm washed and ready – all the mechanics had to wear a shirt and tie and so did I. I go in and get introduced to the foreman. He points over to the corner where there's an old bucket, some bleach and a scrubbing brush, and wants me to clean the toilets. 'Yes sir, no problem,' I say. Everyone was addressed as 'Sir' in those days.

Of course I want to impress and I know from Springfield Court that I'm a good cleaner, so I'm brushing and scrubbing these urinals that are absolutely filthy. No gloves, undiluted bleach, I've got my hands down the toilets and everything. Four hours later the place is absolutely gleaming, though my hands are red-raw from all the bleach. So I go over to the foreman's hut and tell him I'm finished. 'Go back and I'll meet you there,' he tells me, so I go and stand outside, pleased as Punch with myself.

A few minutes later he comes over and takes a good look around. 'You missed a bit,' he tells me.

'Have I? Where?' I ask him.

'Over here.' And he proceeds to unzip his trousers and piss all over the floor – with a big grin on his face.

I'm thirteen and I've got three options: one, cry; two, clean it up; or three, go fucking apeshit.

Unfortunately I did the third.

Picked up a broom handle and whacked him right across the nose with it. He goes down like a sack of shit, blood spurting everywhere and I'm standing over him like Bruce Lee, with this broom handle in my hands! He gets up, screaming and shouting, and we both want to kill each other. We get pulled apart and dragged up to the MD's office.

He's still ranting and raving, wants me dead. The MD, calm

as you like, just says 'Shut up,' and he went completely silent. 'Now, what happened?' he asks.

The foreman kicks off again, saying I hit him for no reason, and I just put my hand up and say, 'No I didn't. I spent five hours cleaning the toilets then, when I told him I'd finished, he pissed on the floor in front of me.'

The MD just looked at him, looked at me, and then back at him. I am shitting it, thinking I've lost the job before it's even started. He then turned to the foreman and said, 'Get your bags and get out. You're a bully, this isn't the first complaint, so fuck off.' He sacked him there and then.

Then I think, right, that's it, now I'm for the high jump.

'And you,' he says to me, 'if you ever raise your hands to anyone again I will deal with you myself. Tomorrow, Sunshine, you're on the workshop floor.'

And that was it. Eight years later and I was still working at that same garage.

Next morning I came in and everyone wanted to make *me* tea, buy me breakfast, because I was the hero that got rid of the foreman they all hated. I was given a mechanic called Ted to work with, and he said, 'Watch, listen and you'll learn properly.' Ted was a fantastic man, the best teacher I could ever have asked for. He taught me things I could only dream of. He taught me the difference between just doing a job and doing a job properly. He taught me methodology: that means not just replacing a component, but finding out why that component failed in the first place. Was it the component that failed or the component that drives that component? I became his shadow, everything he did I did.

As an apprentice after about a year you have to start taking exams, so Ted came up with a test for me which he thought was appropriate. 'Right, Bernie,' he told me. 'I want you to put a new set of contact points in that Silver Shadow over there.'

I'd seen him do it a hundred times, so was confident I could put in the new parts without any help. I went over to the stores, got a new set of points and condenser, put the dressing cover over the wing so I wouldn't have to lean on the car and damage it, took the distributor cap off, took the rotor arm off, and turned the engine's crankshaft round with a socket and a ratchet. The points have to be open on a cam-wheel (a revolving part whose outer edges have projections to activate the points' mechanism). I got the old points open by turning the crankshaft, and in those days it was a twin set of points in there so I did the first set, put them in exactly how Ted had showed me, turned the engine over again until the next set were completely open, took that set out and changed those points too, 'gapped' them with a feeler gauge (adjusted the distance they opened to the correct one), put in a new condenser, and turned the engine over again to check that when the points open they were no more and no less than 14-16 thousands of an inch (known as 'thous').

Job done, Bob's your uncle, so feeling pleased as Punch I went and got Ted. He came across, turned the engine over, checked the gap on one, turned it over again, checked the gap on the other, then looked at me and smiled. Then he smacked me straight round the back of the head, saying, 'Where is the fucking grease on the cam?'

You had a little bit of grease that you put, with a tool like a lollipop stick, onto the cam itself, so that it lubricated the heel of the points as it turned over. And for that I got a clip round the ear. Ted was a perfectionist and thank God he was too 'cos it made me one. I never forgot that bit of grease ever again.

After passing that test I was sent away for a day to

Biggleswade to sit the full exam along with the three other apprentices. We were picked up from the garage at 8 am by coach and taken to what looked like an old school. Inside was a workshop where we spent the morning doing the practical tests. They would put two faults on the car and a guy would stand there with a stopwatch and time how long it took you to find the faults. The first problem they gave me was an engine misfire, while the other one was that every time you beeped the horn the headlights would also flash. There was no diagnostic equipment in those days, not like now, when you just plug in a laptop and wait for it to tell you what the problem is.

So I turn the engine over and sure enough it is misfiring. Now Ted always taught me to look at the most obvious things first – don't try and be too clever, as you'll doubtless be wasting your time. So I looked at the plug leads and sure enough someone had put two of 'em in the wrong order. Sorted that out and moved onto the horn. It was obviously a wiring problem but we didn't have AVO meters (special electrical testing devices), just test equipment, so I made a circuit tester by taking this piece of wire with a bulb on it, connecting it to earth, then connecting the other bulb terminal to a wire, so as to detect any current. Next I took the wires out of the headlight to connect to one side of the bulb. When I pressed the horn the light came on, meaning it had become live when it shouldn't have done, thus proving it was a bad wire. I traced the wire from the headlamp back to the fuse box and there I found someone had taken one of the horn wires off and plugged it into where the headlight cable attached, so it was a simple case of swapping them back again.

I completed the whole job in under eight minutes, while most of the other apprentices were still scratching their heads about the misfire. I was feeling pleased as they sent us off to lunch, until they informed us that in the afternoon it would be the written exam.

Oh shit!

I sit down in the classroom after lunch and we're handed the exam papers. I look at it and it's just a jumble, and my heart sinks. I'm so annoyed with myself that I pinch my wrist so hard I make it bleed. I knew that if they'd asked me the questions verbally I could've told them the answers, but there was no way I could get it down on paper. I literally just sat there for two hours while everyone else scribbled away.

At the end I made sure I was the last one to hand in the paper as we left the classroom. The examiner looked at it and said, 'What's this?' I told him that I was sorry, but I couldn't read and write. He looked at me like I'm some fucking moron. He just couldn't comprehend I'd been sent to take the exam when I couldn't read and write.

He looked down at the exam paper and read me one of the questions and I answered it straight away. The guy just shook his head, didn't know what to do with me. He said he thought I had a learning difficulty as I obviously knew the answers, so I left with a glimmer of hope that he would pass me. A few days later I found out I got 95 per cent on the practical test and 0 per cent on the theory, but they didn't tell me – the results were sent direct to the employers. So I was hauled in front of the MD and I started to cry, just stood there and bawled my fucking eyes out because I was so embarrassed.

An hour later, Ted came over and asked why I hadn't told him I couldn't read and write. Truth was, I was ashamed.

'Right,' he said, 'we're going to get this sorted. Once a week, on a Wednesday, you're going to come home with me after work and stay at my house. My sister is a teacher who works with kids with learning difficulties, and every Wednesday she'll teach you to read and write.'

I don't know if Ted realised it at the time, but I would be staying at his house every Wednesday for the next four years. That's how long it took his sister Melissa, with one-to-one tuition, to teach me to read and write. To help me recognise letters she would make them different sizes, so that A she would write bigger and B she would make smaller, C a bit bigger again and so on. Eventually I learnt the sequence of the letters, so if I had to write the word 'book' I would know that B was a small letter, O was a large letter and K was a large letter too.

To help me with my writing she said: 'Don't look at the page, look at your hand as you write.'

And it worked!

Don't ask me how but it did, and that's how, slowly but surely, I learnt to read and write by the time I was nineteen. Every Wednesday, and the occasional Sunday too, I would stay over at Ted's, Melissa would teach me and June, Ted's wife, would cook the most amazing dinners and I became like one of the family. They never took any payment, they did it for me.

Ted and June had had two children who had died at birth, so they saw me in many ways as their son and treated me as such. I am so fortunate to have had such an incredible family in my life. Ted was the best mechanic I could ever ask to learn from – even though at home he was nice as pie, at work he wasn't afraid to have a go at me and would slap me if I ever did anything wrong or was out of order! And Melissa changed my life by helping me to read and write, and Ted's wife June was like a second mum.

I don't know what I did to deserve those three wonderful people but thank God they came into my life. This would've been a much shorter book if they hadn't!

CHAPTER FIVE

EARLY DAYS IN THE GARAGE

After nearly being booted out on my first day after twatting the foreman, I was with Thomas & Draper for eight years. They were a Rolls-Royce garage but also specialised in all sorts of top-of-the-range cars such as Jaguars and Aston Martins. Basically we worked on the very best cars in the world at that time.

One day, we get a call from Elstree Film Studios saying they've got an Aston Martin that won't start. You'd think that this might be a bit of a glamorous job and a mechanic would jump at the chance to get out of the garage for a while. But this is 1964, everyone's up to their arse in work, we're in Croydon, and Elstree's in fucking Hertfordshire, about two-and-a-half hours away. Basically, it's more hassle than it's worth, so who is told to go and sort it out? The runt of the litter, of course.

So off I set on the bus with my little toolbox for the epic journey up to Borehamwood. By the way, these days I *live* in Borehamwood (I know, posh twat, ain't I?), about two minutes from Elstree Studios. And do they ever call me up now I'm just around the corner? Not fucking once!

I get there a few hours later and after the security guard was satisfied there was nothing funny in my toolbox, he took me over to the studio in question. I've obviously never set foot on a film set before so when I go in it's like another world – what seems like a huge aircraft hangar with hundreds of people buzzing about. All I can remember are lots of lights hanging from the ceiling – it was like a space ship. People are milling around, trying to look busy, but basically all everyone is doing is waiting for Bernie in his boiler suit, steel toe-cap boots, shirt and tie, to turn up and fix the damn car.

As soon as I set foot through the door someone grabs me and takes me over to the silver Aston Martin DB5. I turn it over and sure enough: *vuh-vuh-vuh-vuh*, the thing won't start. So I did the test I'd been taught to do by Ted and found the problem pretty quickly: it was an open circuit on the low-tension side of the coil. I took the wire off, made a new wire for it and a new connector – boom! On the button it starts up and it's purring like a dream.

The director and everyone are over the moon, time is money to them and Bernie's just saved the day. Then this tall geezer comes over to me who I recognise from somewhere but can't think where, and he says, 'Well done, boy,' and slips me a five pound note. Being a young mechanic I didn't see too many of them in those days, I can tell you! I was so chuffed I hot-footed it straight out of there, and didn't stop to think who it was who gave it to me. It was only a few months later when the movie came out and Sean Connery's bloody mug was everywhere that I realised it was Mr James Bond himself, and I'd been on the set of *Goldfinger*!

The only other person I remember from my visit was Harold Sakata, who played the part of Oddjob. He was Japanese, only about five-foot four inches tall but about the same in width as well – he was massive! I certainly wasn't going to forget him in a hurry.

By the time Melissa had finished helping me with my reading and writing I was ready to take the next test in my apprenticeship course. As I was that bit older and more experienced, the test I went for was that much harder. This time for the practical I had to take an engine out of a Rolls-Royce, and I was given 5 hours 20 minutes exactly to do it.

As usual the guys with the stopwatches were there, but they weren't just looking at speed, or how you took the engine out, it was the order in which you did things they were also interested in. Of course, I was confident that Ted had shown me the proper way to do it. First of all you have to make the car safe, which means disconnecting the battery, then you drain the fluids, the antifreeze and the engine oil, to make sure there are no spilt liquids which could make things slippery.

Only then would you start doing the mechanical work. You'd have six boxes, each would be marked with the relevant nuts and bolts, plus some tape so that with every connection you removed you would put some of it on the end of the wire and write on the tape exactly what the part was that it was connected to. You'd take out the wired loom and move it well away from the work area so it was not in the way, then you'd take out the radiator, remove all the hoses, loosen the engine mounts, and then you took all the bolts out of the engine bellhousing, which attached the engine to the gearbox. Then you dropped the exhaust down, got the engine crane to take the weight of the engine, covered the wings of the car so nothing got scratched, and pulled the engine forward and upwards out of the car. You then let the car down off the jacks and put the engine on the bench.

My next test was to reline the brakes on the front and rear. They were asbestos linings so, being aware of how dangerous the fragmented material and dust could be, once you'd taken the brake shoes out of the car's wheel drums, for each one you had to drill out the old lining on the brake shoe, and put a new lining on it, and rivet them to fasten them in place, file them down, and put them on the car – so that when you put the wheel drums back on you wouldn't have to bang them on, but they'd glide on with your hand. If the lining wasn't fitted to the shoe correctly, of course, the wheel drum would be obstructed, and wouldn't fit on smoothly as it should.

Then we broke for lunch, but I couldn't eat anything because I knew what was coming next and I was absolutely bricking it. We sit down in the classroom and we've got one hour to complete the theory test. I turn the papers over and my heart immediately leaps. Where before I just saw squiggles, now I can read it! Where before I sat on my arse looking into space for an hour, this time I completed the whole thing in twenty-five minutes.

If I could explain the feeling of that day, it was a bit like an orgasm that lasted five hours!

When I got back to work and got the results it was massive – it put the company on the map. Nobody had got 100 per cent on both the practical and the theory before as far as I know, and it reflected so well on the company and Ted in particular because it was their training that had got me those marks.

I got a 50 per cent pay rise, a round of applause, an award (I was made Master Technical Engineer) and I was initiated into the Institute of the Motor Industry. So from being kicked out of school at thirteen, by the time I was twenty I had all these letters after my name. My mum and dad cried their fucking eyes out when I told them, and I'd never seen my dad cry before.

It just goes to show that if you're given the right motivation and the right help and the teaching is geared towards the needs of the individual, anyone can achieve anything. I was left on the scrapheap. By the age of seven or eight the teachers at my primary school had given up on me, Mrs Hassinger thought I was a waste of space, I'd been written off as a dunce and insubordinate, as someone who was thick and would never amount to anything.

And I believe that everyone has potential so long as you can find the right work that interests and excites you, and find the right atmosphere in which to learn. I hope that in some small way I'm an example to young mechanics out there who may not have GCSEs coming out of their fucking arse, but want to learn and want to be the best. If they get their head down and put in the effort, they can achieve anything they want.

But just before you think I'm getting a bit too cocky for my own good, I'll tell you a story that brought me back down to earth with a very hard bump.

At the back of the garage was an old Aerial Square Four motorbike. I've never been much of a fan of bikes, I much prefer four wheels to two, but it was transport and so over a few weeks in my spare time I got repairing it. I tested it and all seemed to be running smoothly, so I decided to take it for a run one lunchtime. I pulled up by some traffic lights and a car pulled up next to me so I decided to put it through its paces. As the lights turned green I powered down but the throttle went wide open, the bike shot forward and flipped me right over.

One second I'm sat on the bike at some lights, the next I'm still at the lights but the bike is sat on me! Anyone who's ever ridden one of these old Square Fours knows that they weigh an absolute ton, so not only did I hurt my pride but also got three broken ribs, a broken arm and my left knee was shot to fuck.

I ended up in Hackney Hospital, and my mates from the garage took great delight in coming over to see me and ripped the piss something rotten. After a week or so in hospital I went home but still wasn't able to work. My arm was in plaster but there was nothing they could do with my ribs, I just had to let them heal naturally. When I did finally go back I was on kitchen duty for the first week! I was never a fan of bikes anyway, even less of a fan after that, and I've never once got on a bike since!

I worked at Thomas & Draper Monday-to-Friday, then half-day Saturdays. I'd then head back to Springfield Court and help them out with the taxis on Saturday afternoons and they'd leave me the keys so on Sundays I could do private work. Thomas & Draper didn't mind this as they knew that if they let me do a bit on the side I wouldn't be looking to move elsewhere.

But one day, after eight years at Thomas & Draper, we were sent to a motor trade conference where they told us about all the latest trends in car mechanics and whatever. It was here that I got approached by another garage called Stewart & Arden. I suppose you would say I was 'tapped up'. They knew my reputation and made me an offer I couldn't refuse. Not only did they double my money, but the garage was much closer to home. After all those years of getting three buses to Croydon it was starting to take its toll, so reluctantly I agreed to join them.

A few days later I took Ted out for a drink after work where I told him I was leaving. He was heartbroken, he cried, I cried, we both cried! He'd been so good to me, like a father-figure, and had treated me like a son. But he understood I needed to earn more money and advance my career, so I had his blessing which meant a lot and I knew I could always call on him if I had a problem.

We remained good friends right up to his death in 1982. He was still working in the garage even though he was well into his eighties. He always said he'd never retire, he couldn't afford to, but also what else would he do? It was all he'd ever known and work was his reason for living, he loved his job so why do anything else? In that respect I know I take after him. Here I am now, more than thirty years later, and I'm that silly sod who won't, or can't, retire.

But back at Stewart & Arden I was in for a total culture shock. It was a Jaguar specialist, just like Thomas & Draper, so I was well within my comfort zone as far as the work was concerned, but the atmosphere was totally different. In those days the Jaguar E-type was very popular and there were certain jobs on them that no one wanted to do because they were particularly tricky on those cars – clutches, brake pads and discs and especially head gaskets.

The other mechanics would steer well clear of these particular jobs, so when they found out I enjoyed doing them I was given every bloody car that needed those things doing. However, it worked in my favour eventually because I became so competent at those procedures.

When a car came in the work was assessed and you were given an allotted amount of time to get the job done. So the client would pay for, say, six hours' labour and if you got it done quicker, you took a bonus payment from the fee and the garage kept the rest. Because I was doing so many of these jobs I became so proficient I was finishing them in next to no time, so I was doing ten-hour jobs in six and taking home a tidy bonus, which didn't endear me to the other guys in the garage.

It turned out that leaving Thomas & Draper was a big mistake. There was no camaraderie at Stewart & Arden, and the culture of the place and the reward system meant that no one was willing to help anyone – it was each man for himself. I've always been very free with my knowledge and that's something I am very proud of, but in those days I was probably too eager to help and it backfired. If a mechanic got stuck, rather than go to the foreman for help and lose face, they'd say, 'All right Bern, come and have a look at this,' and I'd help them out. But if I ever needed help I never got the favour returned.

One day a Jaguar E-Type comes in and the foreman says to me, 'Service it and let me know what it needs for the MOT.' I cast my eye over the car underneath, and it's rotten, rotten as a pear (corroded by rust to a dangerous degree). So I tell the foreman, and his words are, 'Just service it only, leave the rest to me, got it Bernie?'

Well, actually I tell him I haven't got it, so he goes on, 'It's a private job of mine, keep schtum if you know what's good for you.'

So I service the car, make a whole list of faults, and this information is torn up by him right in front of me. Then I see him go over to Noel, our MOT tester, speak to him, shout at him, and knock him to the floor. Now, I like Noel, he's a good tester, but I don't want to get involved. All the guys are shit-scared of this foreman, he's huge, but because of what he did to Noel, we all crowd together, and set about him. There's ten of us, and we have difficulty trying to hold him.

The manager comes over and says, 'What the fuck's going on here?'

None of us want to grass anyone up, but as Noel gets up from the floor, he says that the foreman threatened that if he did not write out an MOT he would hurt him bad. The manager asks whose car it is, and the foreman says it's his, and a private job, but that Bernie checked it and serviced it and says it's OK.

'Bollocks!' I reply. 'That car is as rusted as fuck. I gave him a breakdown of all the work required and he tore it up.'

The manager, a quiet man, says, 'Right, that's all I want to hear, I'm calling the police.'

Now this foreman ain't having none of it, and then accuses us all of stealing, doing private jobs, and doing dodgy MOTs (i.e. passing cars that rightfully should fail). That's it, we all try to get near him, we're innocent of what he says, and this big bastard bolts for the door, and legs it. For what? I don't know, but we never see him again.

Once the dust settled I was voted by the guys to be the new foreman. Ah bless 'em, maybe they weren't so bad after all. All goes well for the next six months or so, then shit happens, we are all made redundant.

FUCK!

So I'm back to looking through the paper again for jobs, but I soon found a garage in the East End, perfect for me, that was looking for an experienced mechanic/foreman to run the garage. Little did I know I was going from the frying pan into the fire. I was still in my twenties but had many years' experience by now, so when I phoned the guy up he was very impressed.

This company specialised in engines, transmission and servicing, and they had their own body shop as well. I turned up for the interview and the garage was very clean, usually a good sign. The five mechanics were friendly enough, and the guy was offering me an extra £25 a week on top of what I was already earning – a good pay rise.

As soon as I started something didn't feel right, but I couldn't put my finger on it. Then one day I overheard a conversation where the boss told one of the mechanics, 'Don't worry, just put it through the books and charge him for it.'

I was horrified. I'd never heard anything like it in my life. Where I came from you never cheated someone like that. If you put a new part in then you charged them, but these guys were charging for work and parts that didn't exist, they were blatantly ripping off customers.

So that evening after work I had it out with the boss, confronted him about what he was doing, but he just said, 'If you want to earn the money, you've got to sell stuff.'

I said if the work needed doing I had no problem advising it on the work sheet.

'No,' he says, 'even if it don't need doing, you put it down on the work sheet. We're here to make money and the only way you're gonna make money is by selling stuff.'

I worked the rest of that week, collected my pay packet on the Saturday (we worked half-day Saturdays in those days), walked out and never went back.

Next I spoke to some people in the trade who tell me to go and see a guy called Dave at the arches in north-west London. Now these are railway arches, with big, very high ceilings, and were generally rat infested and had no running water. I got there and asked for Dave – he had heard of me through friends, and he offers me the job of a Foreman Mechanic underneath the arches. The only downside is that they are busy and he needs someone to manage the place totally, and if

I want to earn real money then I will have to start selling the services of the garage as well.

I asked him what he meant by this, and he says, 'If a car comes in for a service, get the mechanic to hand you a list of what needs to be done, then sell the client other work. The more you sell, on top of a standard service, the more commission you will earn.'

Now where have I heard this before?

But I wasn't about to walk out on another job, so armed with my experience, a good work attitude and clients who like me, it's not long before the money is rolling in, the guv is happy, we don't have to rip anyone off and the garage is full to bursting, and for a change I'm content.

And so then comes a Monday morning – pissing with rain, the garage is full, loads of work – when two men walk in and show me their credentials: VAT inspectors. They're followed by a police officer. Oh fuck, what do I do now, I thought?

They go into the office to see Dave, demand the books, but I just get my head down and get on with my work as usual. Three hours later, Dave is handcuffed. As he's marched away he turns to me and says, 'Run the garage, see you soon.'

I hear from Dave's wife that he has got six months for fraud, and so I have a choice: run the garage, keep it profitable, and hope for the best, or leave and be out of another job. So I do the wages, buying of parts, keep all the staff happy, check the work, do the MOTs, bank the takings, and Dave's wife comes in each Saturday to collect cash. But I log how much she takes, keep a list and this is sent to Dave at Brixton Prison weekly by letter, just to safeguard me.

All's well for about four months, when Dave's brother comes in and says to me, 'Sorry son, I'm selling up.'

WHAT?

I ask him why.

'Well, I'll tell ya,' my boss's brother explains, 'Dave will be doing an extra stretch, he's now being done for attempted murder, and the brief reckons he'll do another three years or more. So find yourself another job. I'm sorry, I'm closing it tonight, and I've found a buyer already.'

Well, I'm a cheeky sod, always have been, and my motto is, if you don't ask in this life you don't get – sometimes you ask, and still don't get, but there's always a chance. So I say, 'Can't you have a word with the new owner? Tell him we all want to keep our jobs, and we will run the garage the same?'

He smiles and says, 'He will be here in two hours, ask him yourself, cheeky bollocks.'

The new owner was Sid, and I've seen him around. He was a heavy looking character with a scarred face. I'm introduced to him, give him the overview of what I do, banking, cash, suppliers etc., and he's impressed, I feel he likes me. Then I ask, could I still run the garage for him, with the same staff as well?

'Tell you what son, you got balls,' Sid told me. 'I can see you do a good job, I'm told by Dave's brother that you're honest, and you're not a grass. I'll think about it and let you know on Monday, OK?'

I have a great weekend, but I worry if I'll have a job on Monday. I get up extra early, get to work and park, and as I walk to the garage, I can't believe my eyes.

WHAT THE FUCK?

There's no garage!

It's burnt down, just some cinders and some pieces of steel sticking up at the sides. Oh no, oh my God, what the hell has happened, I wonder? Dave's brother is there, but there's no Sid. There's police and a fire crew there as well.

'What happened?' I ask Dave's brother.

'Don't know, Son,' he says. 'I done the deal, he is the owner now, so it's really not my problem.'

I'm devastated. The garage is gone, all the staff and me are without jobs, and something smells, and it's not down to me. I talk to the fire officer, tell him who I am, and he asks me, 'Who locked up on Friday night?' I don't tell him, as the owner's brother was the last to leave. I have to tell the truth, without grassing on anyone, and Dave's brother has already admitted to the police that he locked up.

The fire officer tells me that it was an oxygen cylinder explosion. The reason why it happened he does not yet know, but investigators are checking it. Now, I did some welding on Friday, but I know for a fact that I turned all the gas off, I'm always double-checking things like that. I don't know what really happened, or what the fallout was, but let's just say that for a few people it was a very convenient 'accident'.

So I'm out of work. Again.

I go to see my bank manager to see if I can get a little of the folding stuff to tide me over. As luck would have it the guy gives me some even better news. He has a client who owns a garage, he's having problems with staff and the receivers are ready to go in, so did I want an introduction to go and see him, so as to try to find out what's wrong with the business?

So I go and meet Ken. He's a real gentleman, whose business was going down through bad management, shit work and clients who were generally, in his words, 'as tight as a crab's arsehole under water.'

This was a garage in Kilburn. I spent two days with Ken, speaking to his service receptionist, mechanics and bookkeeper, and I knew that something was not quite kosher. Ken employed me as a foreman in charge and told me that he would pop in twice a week as he has other businesses to run.

After my first day on my own, my suspicions were confirmed – the banking did not match the money that came in. Two of the cars that were repaired previously were back in the workshop, so I looked at the invoice of the previous jobs and went over with the mechanics what had been done. The invoice stated, to rectify a misfire, 'removed and replaced head gasket, road test, no further fault found.' After checking the car, it was obvious that the head gasket was not removed but just cleaned, and the mechanic who did this job denied doing a head gasket, although it was billed for on the client's invoice.

I'm not an investigator, but neither am I a prat. And if the mechanic denies doing a head gasket and the client has been invoiced for it, then who wrote the fucking bill out? Inspector Bernie Fineman investigates...

I notice in the office a locked metal filing cabinet beside the reception desk. I ask the receptionist, Paula, what is in the cabinet.

'Oh,' she states, 'this does not concern you, it is purely for the accountant.'

'No,' I say, 'I wish to see what's in there.'

'I don't have the keys,' she says.

So after everyone went home for the day, I called Ken and told him that I needed to see what's in the cabinet. Something is wrong, but I don't know what. He says that he does not have the key, Paula has it. Now my hackles are raised and suspicions confirmed, so I pick the lock.

Lo and behold, inside are hundreds of invoices, neatly filed A-Z, so I pull out the invoice for the so-called replaced head gasket, and the client's invoice states £1,050 but on the invoice in the filing cabinet it states £196. So my thinking is as follows: the mechanic does the job, sends Paula the invoice for £196, Paula then re-invoices the client for £1,050 and pockets the balance (850 quid).

I need an urgent meeting with Ken, I need to find out how long this has been going on for and how long has Paula been employed at the company. Whilst looking at other invoices, ranging from £800–£2,000, I am wondering how many other thousands she has creamed off from the company. What I need to do now is go over all the parts invoiced and establish what has or has not been put onto the cars that have been repaired.

After my meeting with Ken it was decided not to call in the police, for reasons that only Ken knows about, but Paula was sacked on the spot. How long this scam had been going on for, God only knows.

After a few months the business is back on its feet, it is making a profit, and I am loving what I am doing. Then something terrible happens.

It was a dark Monday, bleak out but we're busy, and the bodyworker, Tom, is fitting a new chassis outrigger to a Jaguar which I failed on an MOT, all the mechanics are busy, and I'm talking to a client. I hear noises, like thunder, look over to the ramps and see flames coming out of the Jaguar that Tom is working on. I'm shouting to him, but he can't hear me. I run over to the ramp and propel myself at Tom, and, just at that moment, the tank explodes. I'm covered in flames, Tom's on the floor, writhing around, acetylene torch in hand, but I'm lucky – all the mechanics grab fire extinguishers and put the flames out.

I'm burnt, not so badly, just my arms, eyebrows, and my moustache, my pride and joy – and what hair I had in those days was gone. The worst were my arms, they hurt like fuck, but the ambulance crew does a good job on me. Tom, the silly fucker, has his arms broken where I propelled myself at him. I heal quick, and forget, but safety will always be my prime thought, even though it's not everyone's. By the way, this was not the last time I saw someone burnt by carelessness...

I was with Ken for fourteen months after that accident before it was time to move on again. I got a call from a mutual friend that he knew someone with a breaker's yard and a garage attached in London. I go and see the guy, he's heard of me, and takes me on, cash in hand, I pay my own tax, no questions asked.

It's a series of lockups, I'm not allowed to do invoices, just have to repair the cars, and occasionally remove second-hand bits for clients from the cars in the breaker's yard. I'm earning, it's hard work, no ramps, we had to make do with just jacks, and there's concrete floors, but I'm working.

An MG Maestro is towed in, it's been flooded with water, so the electrician is called to check it out. He does all the normal stuff: checks the ignition amplifier, removes the plugs and leads and powers it up. He asks one of the boys to turn over the starter so he can check for spark, and that's when disaster strikes. The engine turns over, fuel is shot out from the plug holes, the leads ignite the fuel – and he's alight, from head to toe, and screaming a blood-curdling scream.

The other mechanics come over, they are shell-shocked, can't move, fear and fascination together paralyzing them. I see this, jump the stairs from the office, grab a hose and pull him to the floor. I'm dousing him with water, trying to put out the fire which is spreading all over his body and face. 'CALL AN AMBULANCE!' I shout, 'FUCK SAKE SOMEONE HELP ME!' My mind runs riot, remembering all the pain I felt when it happened to me, and I'm in near-panic mode.

The flames are out, but this poor fucker is in agony, his

clothes literally melted onto him, and his face, oh my God, it's red raw. He can't breathe, and smoke from the fire has filled his lungs. I remember something from my old days when I did St John's Ambulance training, so I try chest compression, but there's nothing, so I breathe into his mouth, wait for the fall and rise of the chest and lungs, but nothing.

So I take drastic action. I proceed to slowly cut into his windpipe above the breastbone and the soft tissue at the base with a Stanley blade and open his airways with a biro outer casing from my pocket. I hear gurgling, and coughing, but he's breathing now.

Where the fuck I got the balls to do this, I will never know. The ambulance arrives and the paramedics commend me on saving his life. I'm a hero, then I puke my heart out.

I went to see him at Mount Vernon hospital three times a week for two months. He had all sorts of operations to rebuild his face and arms, poor bastard. If only he had taken the coil wire out none of this would have happened, but then it's easy to talk after the event. We are still mates today, and I'm delighted to say that the operations have been successful. His girlfriend stood by him. What a marvellous lady she is, bless her.

CHAPTER SIX

THE BLIND BEGGAR

As my name got around as someone reliable, people would come to me directly to work on their cars, which I'd do in my own time. By the time I was eighteen or nineteen I was getting so much private work I decided to rent my own lock-up garage so I could come and go as I pleased and work on the cars.

Through a friend of a friend I found a place on Bodney Road in Hackney where a guy had some garages. The rent was pretty reasonable, four pounds a week, which I could earn doing one job, so I took it. Word soon got around that there's this guy on Bodney Road who does work cheap but does a good job, and I'm getting approached by all sorts of people.

One night I'm in the Blind Beggar pub on Whitechapel Road and this guy called Al approaches me about doing some work. He's a huge guy, looked like Lurch from *The Addams Family*, a real hard bastard but smart as you like in a black suit, white shirt and black tie, with Brylcreemed (smoothed down with hair-grease) hair. He had a scar running down from his left eye to his chin. When I asked my old mate from school Jimmy Nunn what Al did he said, 'He collects debts, and if they don't pay he kills them.' Well, I was too scared to say no! We were in the Blind Beggar after all, and I knew the reputation that some of its patrons had.

Over the years there I met the Lambrianou brothers, George Cornell, Mad Frankie Fraser – all the big names in the East End at that time. You'd think it was a dangerous place to be, but in fact it was anything but. It was a regular boozer, always clean and immaculately presented, and the atmosphere was always welcoming – so long as they knew you.

If they didn't recognise you or they didn't like the look of you it would be: 'Oi, fuck off!' It was a jolly place, lots of laughing and joking and they'd have a musical-comedy turn every Saturday night, and there'd be sing-songs around the Old Joanna (piano) when the mums and dads would have a knees-up. Across the road was a fish 'n' chip shop, so you'd go over and get your fish 'n' chips, bring 'em back, have a pint, meet your mates and have a great night out.

Most of the guys there had bent noses and cauliflower ears from boxing, but were always smart and always wearing suits, and the women looked as hard as the men half the time! But if your face fitted and they knew you, you always got a warm welcome. You could walk in with no money and there'd always be someone more than happy to buy you a pint.

I suppose because of its reputation you were either on the inside or on the outside. If you were on the inside you were part of the family, and it was a big happy family, but if you were on the outside you knew to stay well clear of the place. So as a result of that there was never any trouble – not like in pubs today when you only have to look at someone the wrong way and you'll get a bottle across your face.

For instance everybody had manners. If you bumped into someone it was, 'Ooh, sorry about that, let me get you another drink.' There was no shouting and screaming, it was respectful. Everyone was there for a good time and that's what they had, it was like a second lounge with all your mates there. It was Jimmy Nunn who first introduced me to the Blind Beggar. His dad had been drinking there for years and so we had no problems, and it was Jimmy who introduced me to Al (never Alan), who became one of my best customers for private work.

I wasn't naïve, I knew the kinds of people who owned the cars that Al brought to me, but my philosophy was 'ask no questions, tell no lies'. And anyway, nothing was ever said but you knew you were far better off doing the work than not, and it was a brave man in those days who said 'no' to people like Al.

Because of the clients I now had, my garage was full of Jags, Rolls-Royces, old MGs, a Triumph Mayflower, all this type of thing. One day the landlord came by, saw all this and said, 'You're obviously doing well for yourself, I'm going to have to put the rent up.'

Later that week I was returning one of the cars to the Blind Beggar and I said to Al, 'I'm sorry, it's going to be tricky for me to do work in the future, the landlord has put my rent up so much I can't afford to keep it.'

Al was very disappointed.

The next weekend I was working in the garage when the landlord came up to me, a bit sheepish, and said, 'There's no more rent. Don't worry about it, you don't owe me anything.'

I was flabbergasted, but what a result! I worked out later

that Al or one of his guys must've gone to see my landlord. He had several properties in the area and was always having trouble getting rent out of people. Being an expert debt collector, Al obviously did a deal with him, making sure he always got his money on time, so long as he kept Bernie's garage open.

A few months went by, I was rent free by now and there was a steady stream of work. Then one Friday night I get a telephone call at about one in the morning. I recognise the voice and he says, 'Remember me, I did you a favour.'

Now I realised that Al didn't just sort out my landlord to keep my business going, he did it so that I was in his debt and the law of the East End says that he can call in that debt any time, and that time was 2 am tonight. All I have to do is collect a car from Surrey Docks at 2 am, ask for the guard by name, and tell him Tom sent me, take the car to south London at the address he gave me, and we're quits.

'It will be done,' I say, and the call ends. I arrive at Surrey Docks, ask for the guard by name, he hands me a parcel, keys for the car, and I winch it up on the breakdown truck and get on my way to south London. All goes well, until I'm at Camberwell Green at about 4 am, and the Old Bill stop me and they ask me questions. I tell them my name, and who I work for, and explain some cock and bull story that I have to deliver the car before the client goes to work. They're happy with the story, so with me sweating like a pig, I drive to my destination and tear back home, just in time to wash and go to work. I'm shattered, but relieved the job's done and the favour has been returned.

A week or so later I took some more cars back to the Blind Beggar. Al approaches me and I'm shitting it. He says, 'Son, you done a good job, you never grassed or gave out names, take a ton, I can now trust you, you're one of us, any problems you get, keep it in the family, tell me, and it's sorted.' Al handed over a hundred pounds in cash.

RESULT!

I regularly took cars that were fixed back to gangland haunts, including Esmeralda's Barn, the Krays' nightclub in Knightsbridge, and the Amherst Club, as well as various clubs and pubs in London's East End. I met Diana Dors and her husband Alan Lake, and Barbara Windsor and Ronnie Knight. In fact, a few years later I moved into a flat across the road from Barbara in Hendon and she came to my place for dinner. A wonderful woman, great company.

One day I returned a car to the Blind Beggar, a Mk 10 Jaguar. I went in and saw Al, he asked me how much he owed me, I told him, and he gave me the money, no questions asked. Then he said, 'There's a couple of people I want you to meet. This is Ronnie and this is Reggie; gentlemen, this is Bernie, the guy we use for some of our cars.' They were both incredibly polite, said they were very pleased to meet me and that they'd make sure that any work came my way. They insisted I have a drink with them, which I could hardly refuse, so they ordered me a double whisky.

Now, anyone who knows me knows I'm not much of a drinker, and whisky I cannot drink. I don't know why, but there's something about whisky that brings something like a cloud over me, I suddenly feel very angry when I drink it. So be warned, never buy me a whisky, it turns me into the Incredible Hulk! But on this occasion, I could hardly say no, so I accepted it and drank it politely and tried not to conceal how tipsy I was feeling.

Then Reggie put his hand in his top pocket and pulled out one of the big old pound notes and handed it to me, saying,

'Get your mum some flowers and your dad some tobacco.' From that day on whenever I returned a car to the Blind Beggar, whoever it was I saw would always give me an extra pound to buy flowers for my mum and tobacco for my dad, and the Krays would always make sure I got a lift home.

Other people I did work for around this time included Jack 'The Hat' McVitie, who was a real character. He was losing his hair so would always be wearing different hats, hence the name, and he was always pissed! He wasn't a hard man like the others – he was a ducker and diver, a wheeler-dealer who'd rip you off any way he could to make some money. But he never ripped me off, he knew I did a good job for him so kept me onside. There was one time when I returned the car and Jack didn't have any money to pay me, so Jimmy Nunn laid it out for him and said, 'Right, you owe me that money now, and I want it tomorrow.' And, sure enough, Jack had the money the next day for him. He was that sort of character – someone who could find money when he needed to.

Another person who I always thought was a really nice guy and always got on very well with was Frankie Fraser, who I did several cars for. We were never friends, it was a business relationship, but he always paid up, shook your hand and was very polite, and a pleasure to do business with. At least he was for me, though there may be others no longer with us who might disagree. One thing he always said to me was, 'Never buy any jewellery. If you want something for your mum or your missus, you come and see me.' That was his speciality, smash-and-grab from jewellery stores, though I never took him up on his kind offer.

I was growing up. It took me a long time, what with discos, girls, snogging, the occasional bit of rumpy pumpy in the back of my new car (well a five-year-old Ford Zodiac, auto 6 cylinder).

Then shit happened.

It must've been late '67, early 1968, and my landlord at the lock-up had been forced to sell so I had to move out. Via a friend of a friend I'd got speaking to the owners of a garage in Hackney and they let me lease some space from them to continue my out-of-hours work. I'm hard at it one Saturday morning, quite early, when a load of police cars pull up outside, about four of them. The coppers run into the garage, no messing about, shouting, 'You're wanted for questioning, you're all coming with us, now!'

Before we know what's going on we're all thrown into the backs of cars and transported to Stoke Newington Police Station, where we are interrogated. I was put in the back of a car with my mate Tony, and we were protesting, 'What the fuck's this about? What's going on?' The coppers in the front just said, you'll see when we get there. Tony and I didn't say a word the whole way there, we knew better than to open our mouths. But in fact there was nothing to say, we'd not done anything wrong, all I could think was there must've been some dodgy stuff going on, fiddling the books or whatever.

There was a new inspector in town, 'Nipper' Read, who thought of himself as the big cheese and was determined to get the Krays locked up any way he could. Ronnie had shot dead George Cornell in 1966 and Reggie was said to have murdered Jack The Hat, but the Lambrianous got sent down for it instead. Witnesses were too scared to come forward, the Krays seemed untouchable, but 'Nipper' Read had other ideas and ran a campaign of pulling in just about everybody in the East End, including me.

We were all taken to separate rooms. They told me to sit

down and asked if I wanted a cup of tea. Yes please! I was interviewed by a detective sergeant who said he knew I'd done work for the Krays and asked me what I saw in the back of the cars. 'The boot,' I replied, and his words to me were, 'Listen you cocky little fucker, this is a murder investigation, keep your smartarse remarks to yourself and answer the questions.'

I then said, NO, I saw nothing but a toolkit and carpet in the boot. He asked me if I saw any guns, knives or axes, or blood, and I said NO, I am just a mechanic. So he goes on to ask, did you receive any cars damaged with blood on the bumpers? NO I replied, I am just a mechanic. He kept on, with words to the effect: 'I know you know something, what are you withholding?'

'NOTHING,' I replied. 'I am a mechanic, I do my job and do not take notice of anything else.'

He then asks, 'Are you being told not to say anything? Is someone paying you to keep your mouth shut?'

'NO,' I replied, 'I don't fear anyone, I do my job and I don't hear or see anything else.'

After twenty minutes of this, realising I wasn't going to crack, he told me to piss off before he changed his mind.

I'm shaken, but all I can think about now is: How am I going to get back to work? I'm in my boiler suit, my wallet was in my civvies (clothes that were back in the garage), so I ask the desk sergeant if I can get a lift. He just looks at me and tells me to fuck off. I'm the first one out so I have no choice but to start walking. Eventually the other guys are slowly released and we catch each other up and make the long walk back to Hackney in our steel toe-capped boots.

Was it the truth? Was I hiding anything? Course I fucking was. We all knew what was going on but business was business and these people weren't just loyal customers, they were dangerous ones too, so you just got on with it and kept your mouth shut. If you did see anything you not only didn't tell the police, you didn't tell the customer either. If they got a whiff you might blag you'd be well in trouble.

The truth is I saw sawn-off shotguns, and I saw cars with blood on the grill. Because I was on call for them 24/7 I would get called day and night, and told there'd been an accident and a car needs moving. Even if it was three in the morning I didn't hang around because I knew those guys needed it moving a.s.a.p.

As an example of that kind of call, I remember once I had a request to attend to a 2.4 Jaguar which had hit a lamppost outside the Super Cinema, Stamford Hill. They said it had gone out of control, the keys were on the top of the back wheel, and they put the phone down. When I got there I couldn't believe it – it was a total wreck. The Jag had hit the lamppost so hard it was like in a V-shape around it and the lamppost was almost touching the windscreen, with the engine virtually in the driver's compartment. They must've been doing a hell of a speed to do that much damage, so I could only assume they were chasing someone.

That's the only reason you'd do that speed on that road, and of course the guys in the Jaguar couldn't have been the ones being chased because they would've soon been caught after the accident. But as always, I asked no questions. I got it on the recovery truck, drove back to the garage, reversed it into the yard and locked up. One of the local scrapyards used to do pretty well too out of the Blind Beggar because that Jag and many others ended up there.

Eventually, 'Nipper' Read got his men. Ronnie and Reggie, along with fourteen others from their firm, were sent down.

What with them in prison, and with Jack the Hat and George Cornell dead, that meant bad news for business for me. I still did work for some of the locals who weren't involved in the firm, but eventually the jobs from the Blind Beggar petered out, and it was never the same again.

CHAPTER SEVEN

THE LADIES IN
MY LIFE

I was introduced to my first wife, Jacqueline, through a friend. I was a grease-monkey mechanic and she was from quite a well-to-do area called Kingsbury in north-west London. I'd never been to bloody Kingsbury in my life, and thought I needed a passport to get there. Anyway, we hit it off straight away but her parents never liked me and I suffered them. We were nineteen or twenty and in those days when you got with a girl you were expected to get married at that age. So people would ask us, 'Are you getting married?' We'd say yes and before we knew it we were! But as I said, her parents never liked me and I wonder if that was my attraction for her.

We were young, we got married, but we probably shouldn't have done. Sure enough, in time we just began to grow apart. I was working long hours to keep the little house we had, wasn't used to having responsibilities and we just drifted away from each other.

For a while we tried to keep our marriage together and

thought the way to do that was to have a baby. We'd been trying for a while but it didn't happen, so I had some tests and it turned out I had a very low sperm count so was told I couldn't have children. Fortunately we had some very good friends who had a friend whose family wouldn't accept her back because she was pregnant out of wedlock. We were talking and the fact that we couldn't have kids cropped up, and so they asked if we'd be interested in adopting. So we met the girl and she was very nice and was happy for us to adopt her baby. We spoke to our solicitor, did everything legally, and I was actually there when she gave birth. So she's my eldest daughter, Lisa Marie, a wonderful girl and a fantastic mother, who's given me two beautiful grandchildren called Daisy and Mia.

Of course, having a baby didn't save my marriage. I was with Jacqueline for twelve years and we finally divorced when I was thirty-two. I spent the next six years making up for lost time and sowing my wild oats, and then I met Lisa Blume – at the Middlesex & Herts Country Club of all places. She was like the 'Geezer Girl', wouldn't put up with any of my shit, called a spade a spade, she just wouldn't take any fucking nonsense.

Lisa was working for a car dealership, on reception, so she knew how to handle blokes. I knew I liked her, but she was sort of seeing this other geezer who kept pestering her. So one day we were up the Country Club and I got him by the throat and he never pestered her again. After a few months of knowing her, we were getting on like a house on fire, and my mates booked a lads' holiday to Thailand. Lovely-jubbly I thought!

But Lisa found out about it and said, 'You go to Thailand and that's it, you can forget about me, Sunshine.'

So I had to cancel the plans with my mates and I went with her instead. A few months in and I'm already taking this girl on holiday to Thailand, in a beautiful hotel overlooking Pattaya beach, not an everyday destination back in the early 1980s. I must've loved her!

It happened to be her birthday while we were there, so I booked us into this beautiful hotel restaurant as a treat and ordered a birthday cake without her knowing. She's always been slim, my missus, a great figure, no matter how much she eats. But when this cake was brought out it was massive! I was expecting a Lyons Corner House job but this thing was fucking huge! And she's shovelling it away, getting right stuck into it. I'm sat there watching, stunned, and I just jokingly said, 'If you think I'm going to marry a fat lump you've got another thing coming.'

And she said, 'You've proposed!'

'What? No I didn't.'

'Yes you did, you proposed!'

Next thing I know, when we got back to England all of her family, the bloody 'Blume Mafia', were there to greet me. I thought 'I've got to put a stop to this', so she went back to her parents' place and I went to my little bachelor flat in Hendon, made a few phone calls to the lads to arrange a night out, and thought, 'Great, I'm out of here.'

Three hours later I get a knock at the door. It's Lisa with her suitcases. She moved in there and then. And we've been together ever since! Lisa definitely helps keep me young, she's fifteen years younger than me. When I told the guys in the garage I was getting married one of them said, 'Why don't you marry someone your own age?'

I said, 'I can't, they're all dead!'

When we were first married money was tight. We lived in

my little flat in Hendon but wanted a place that was *ours* rather than *mine*. At the garage we had a 24-hour breakdown service and recovery truck, so I speak to the guv, tell him I want to be on 24-hour call, and so he lets me take the truck home with me. We agree to do a 60/40 split with any breakdowns and if it can't be fixed roadside, a good job will be a 40 per cent bonus if the car is brought to the garage for repair.

In seven days I did just ONE breakdown, so I decided to be a 'Motorway Pirate'. This is a lonely and risky job, cruising up and down the M1 motorway looking for breakdowns. Now I realised that this ain't no way to start a marriage, so I restricted my on-call times to three nights per week. I got some small jobs: tyre changes, run-out-of-fuel gigs, and tow-off-the-motorway jobs, plus the occasional accident recovery.

I shared the profits with the guv, and to be frank it brought in a small yet good addition to our household, but 'Her Indoors' was getting pissed off with me working night and day, and it affected my health as well, what with only getting an hour or two's kip some nights. So I gave it up after a month or two. But it was a real experience, cruising on the open road, not knowing when the next job might turn up, living by your wits and the luck of being in the right place at the right time.

Eventually Lisa and I saved enough to move into a small house in Hertfordshire, and did all the decorating ourselves, and tried to make the place look homely. One Sunday we have an almighty row, as couples do. Now Lisa is a sweet lady, but believe me you don't wanna mess with her. It goes on and on and on and on, until she states: 'THAT'S ALL YOU WANT ME FOR – COOKING, CLEANING AND FUCKING!' And with that she strips off all her clothes, runs around the room, then out into the garden and carries it on, ranting and raving.

Mr Cool me, I just lock the patio door, and she is still naked – I bet the neighbours loved it. I refused to let her in until she calmed down. After that peace was restored and we both saw the funny side and laughed. Lisa has to have a sense of humour, living with me.

I've been married to my wife Lisa for thirty years now. She's the only person on earth I'd ever let talk down to me, I just shut up and take it. Nadine, our daughter, is twenty-eight now, bless her, and married with a beautiful baby girl called Chloe – I could pick her up and eat her! Here's one quick story that involves Nadine, and is possibly the most extraordinary coincidence I have ever known.

When I was a lad we used to all go to the Coronet Club in St John's Wood (now a very swish café/restaurant called Richoux) on a Saturday night for a bit of dancing and how's your father. In those days it was two shillings and sixpence (15p) to get in, and you got a free record! One night we all squeezed into my mate's father's car and my ailing Triumph Herald, registration number YMC 600. We looked hot, all bathed, Old Spice aftershave and an itch between the legs for some action. We arrived, paid, and searched around for some likely candidates. I spotted a good 'un: Mary Quant-type hairstyle (a very fashionable look back then), all made up and *Mmmm* she had lovely legs.

'What's your name darling?' I asked her.

'Sandy,' she replied.

'Hello Sandy, would you like to dance?'

'OK then.'

We took to the floor like Fred and Ginger (Astaire and Rogers – a film star couple, famous for their accomplished dancing onscreen) and 'twisted' to Chubby Checker – a sixties pop star who invented his own dance of that name. Obviously

I'm hoping to impress and maybe get back to hers or mine for a bit of rumpy-pumpy, so I ask Sandy if I can give her a lift home. She enthusiastically agrees but it was only a bit later on that I found out why. She lived in fucking Cornwall!

Being a geezer I could hardly change my mind – and I did really fancy her – so I stupidly said, 'No problem.' Except there was a problem. I only had my Triumph Herald and in those days Cornwall was a seven-hour drive away. But she offered to pay for the petrol so I took her home.

When we got there I was knackered and in no fit state for any physical exertion, so I slept on the couch and met her mum and dad when they got up the next morning. I think they were rather impressed with my selfless offer to see their daughter home safely so we got on quite well.

We all went for a walk and then sat down for a nice Sunday lunch before I set off home again at 5 pm. I got home at midnight and was up again a few hours later to get to work at Thomas & Draper. Of course, everyone thought I was a fucking idiot when I told them I'd been to Cornwall and back in the last twenty-four hours! But it was worth it in the end, as Sandy and I continued a long-distance relationship for the next few months, taking it in turns either for her to come up to London or for me to head down to Cornwall.

The friend of hers that she was visiting when she was in the Coronet Club then got engaged to my mate Rod, it was the 'thing to do' in those days. So I said, 'Why don't we get engaged too?' And she said yes!

When I told my dad he said, 'Son, you're nineteen, you're hardly earning any money and you've only got a battered old Triumph Herald. Rather than get married why don't you spend the money on buying a new car instead?'

My dad spoke a lot of sense, so after thinking this over for a couple of days I broke off the engagement to Sandy and traded my Herald in for a Sunbeam Alpine!

At the time one of my famous clients was a singer called Eden Kane who'd had a number one hit with 'Well I Ask You' and his brother was also a singer called Peter Sarstedt. Anyway, Eden had a blue Sunbeam Alpine that I was working on at the time and I absolutely fell in love with it, loved it more than Sandy in fact, so that's why I got the car!

I never gave Sandy much of a second thought after that, until about forty years later when Nadine brought a chap home. She was in her early twenties and this guy was a bit older, probably late twenties, and had only just split up from his wife. At some point early on in the relationship he'd obviously told his mum he was seeing a new girl and said her name was Nadine Fineman.

'Fineman? What's her dad's name?' his mum asked. Well, you can only imagine Sandy's face when her son told her it was Bernie. Probably she was just as astounded as I'd been when he and Nadine told me who his mum was.

Fuck me! You could've knocked me down with a feather!

Turns out that Sandy eventually left Cornwall for good and moved to London, got married and had a family. But still, what are the chances of two people who live at opposite ends of the country having a teenage romance and forty years later their children coincidentally meeting and dating as well? To this day it still freaks me out when I think about it.

That was the first time Nadine nearly gave me a heart attack and she's done so a couple of times since too. When she passed her driving test I bought her a VW Polo, serviced it, put alloys on it, blinged it up. It was her pride and joy.

A few weeks later me and Lisa are out for dinner with our good friends Phil and Sharon when we get a phone call to say that Nadine has been in an accident. There's five of them in the car, she's gone round Mill Hill roundabout, lost control, clipped the kerb, and the car's flipped upside down and skidded twenty feet on its roof and hit a tree. The motor is barely recognisable as a vehicle but all five of them walked away from it without a scratch, not even a bruise – an absolute miracle.

The other time she had an accident was when she was driving home from her job in a Bureau de Change at City Airport one night along Allum Lane, a steep and narrow road, and a white van veered over into the wrong lane and hit her head-on. Again, her car was a total write-off, about two feet shorter than it had started out, but miraculously, still Nadine walked away without a scratch.

Someone, somewhere is clearly looking down on her but I hope to God she doesn't have another incident, otherwise it could be third time unlucky. You remember I said how my mum's hair went completely white when she heard about my dad's car accident? Well, I don't have any hair, but if I did have, God knows what colour it would have been after Nadine's exploits!

So I've got a wife, two daughters and three granddaughters: I'm surrounded by women!

No wonder I'm bald and stark raving fucking mad!

CHAPTER EIGHT
GOING PLACES

The motor trade is a relatively small industry, so you can build a reputation quite quickly, but you can always lose it just as fast. By the late seventies I was making a good name for myself so if I lost a job, with a few phone calls on a Saturday and Sunday, come the Monday I was back in work.

I had heard there was a vacancy at the Ladbroke Grove Motor Company. It was run by a couple of brothers, the garage was spotlessly clean and they did repairs for cars that were more your everyday Fords and whatever, so when I joined – and with my client base – they started getting the Jags and Astons in, which they were well chuffed about.

Back in the late seventies, Ladbroke Grove had a large immigrant community, as it still does now. The thing about garages is that we see all walks of life coming in, for if you've got a car it doesn't matter who you are, at some point you're probably going to have to take it to a garage.

So where in the East End I had some loyal customers from

the Blind Beggar, in Ladbroke Grove it was the Yardie boys. These were gangsters of sorts, but the new breed. In the East End in the sixties it wasn't about drugs, it was about protection rackets, extortion and robberies. In the West End it was the Caribbean immigrant crew pushing drugs and no doubt a very different operation, but they had many of the same values as the old East End gangsters. They were always polite and respectful and if you looked after them they looked after you.

These were big, hard guys, but had clearly never done a day's work in their lives yet were still able to drive around in blinged-up BMWs. Go figure. So as they drove round Ladbroke Grove they saw all the nice cars that we were now bringing in through my contacts so they started bringing their pride and joys in too. Andy, one of the owners, was very wary of them but, me being me, I get on with everyone, so I got speaking to them and over time built up their trust and so soon enough they were bringing nearly all their business our way.

These were all the 'Faces' in the area, the guys who ran things. Eventually, one of them said, 'You'll never have any problems round here.' Lots of businesses were getting broken into but once the word got out that they were bringing their cars to us we never had a moment's trouble.

The rule with dealing with guys like this was to keep your eyes and ears open but your mouth shut. Payment was always done in cash, they would have bagfuls of notes, and they would always give us a little extra to buy a drink. I remember one time one of the big boys needed some major work done on a Cadillac and when I finished I got a £50 tip – not bad at all in those days, and at Christmas it was bottles of champagne.

The main man was a guy called, I won't say his real name,

let's call him 'K'. He was a man-mountain, obviously worked out a lot, with long dreadlocks down his back, immaculately dressed in charcoal grey suits, grey shirt and red tie, gold chains and rings, plenty of bling, always wearing hats, even Stetsons! His watch was so big it looked like he had Big Ben on his wrist.

K would be chauffeur-driven everywhere and virtually every time he came in he had a different car, and sat in the back with smoke pouring out from a big fat joint. He was like the Jamaican Godfather. K would always say the same thing. If I did a good job he would say, 'Bernie bwoy, dis ting is blouse and skirts!'

Blouse and skirts? I'm used to Cockney rhyming slang, and it can be odd at times, but I could never get my head around 'blouse and skirts' but I took it to mean I'd done a good job.

Occasionally you'd get a car in with a particularly big dent in the wing. Whether it was a football that had hit it or someone's head I never knew. As before, it was 'ask no questions, tell no lies'. Sometimes they'd say specifically: 'Don't go in the boot' and you wouldn't, not even out of curiosity – it was much better not to know.

They had money and they loved their cars, those boys. K had a Cadillac Seville with a full leopard-skin interior, even the ceiling, carpets, and steering wheel. In fact the only thing that wasn't covered in leopard-skin were the furry dice hanging off the rear-view mirror. The other cars I particularly remember were his Chevy Monte Carlo and Corvette Stingray, you didn't see too many of them around. He had different cars all the time and one day I said to him, 'What, have you got your own garage?'

'No maan, I just like me cars.'

Around this time a mate of mine thought he'd spotted a gap

in the market when it came to the Corvette Stingray. They were coveted cars but very expensive to buy in the UK, and of course they were left-hand drive. He was importing them and getting them converted once they were here, but he wasn't happy with the quality of the cars that were being sent, and the expense of getting any parts sent over was cutting into his profit margins. So, one day he says he wants to go out to Miami and source the cars himself and would I go with him to check they're all kosher and do the conversions out there? He tells me he'll split the profits with me.

It's a tempting offer but I said I needed a wage, I need to live. In a flash he slams five hundred quid into my hand and says, 'There's plenty more where that came from. You be the mechanic, I'll be the businessman, we're all going to earn some folding.'

So he was going to pay me hand-over-fist to leave the dirty East End and move to Miami. They say that if an offer comes along that sounds too good to be true, it probably is. Well, this was the exception to the rule, it sounded good and it turned out to be even better!

Seven days later I'm in Miami. Wow, what a place! The sun was always shining, the women had long legs, nice round arses and an accent I didn't understand. Fortunately, they love mine! I've never seen as much action before or since as I did in those months in Miami.

But back to the cars for a moment. We go to see some Corvettes and check them out, but of five we see only one that is any good, and it's very expensive. We walk along Biscayne Boulevard feeling a bit dejected, and I hear someone from within a garage shouting 'Motherfucking English crap!'

Curious now, I venture into the garage, and see this so-called mechanic trying to get a rear axle shaft out of a MGA

1600, and he's thumping the axle casing with a massive club hammer. I step inside, he turns round, faces me and says, 'Who are you, what do you want?'

I reply, 'Sorry, I heard you swearing, "fucking English cars", so I just had to get a look at what it's about. I'm a mechanic, so would you like me to show you an easy way to get the axle shaft out?'

'Go ahead,' he says, 'I'll get you a beer.'

By the time he comes back with the beer, the axle shaft is out. 'What the...?' he gasps 'How ya do that?'

Just then the MD walks in, his name is Giancarlo Seri, and he asks me who I am. The mechanic tells him he had problems with this car and this here guy helped him. Giancarlo offers me a job on the spot but I tell him I'm here with a mate, we want to buy Corvettes and ship them to the UK, then convert them to right hand drive, and make a good profit, as there is a big demand there.

He says, 'Well, you take the job, I will get you the best price Corvettes, convert them here and ship them to the UK, you can earn good money, an apartment, and a car.' Fuck me, these Yanks move fast! And it didn't stop there. Within a couple of weeks I have my green card so I can work in Miami legit.

Soon I have a small apartment right near Key Biscayne, I show Bill (my English car-buying mate) some trade secrets, and then I introduce Giancarlo to my UK suppliers for English car parts, and he ships them out to Miami at a third of the price. We start getting very busy, particularly with Jaguar XK 150s, 3.8s, Rolls-Royces and Bentleys. I'm earning double what I earn in the UK, meet some great people, but I wonder how Giancarlo got me a work permit and green card so quickly. I feel he's 'connected'.

This is confirmed when he introduces me to his brother, who turns out to be the chief of police. Now, this brother drives a Jaguar 3.8, a $50,000 car, not bad on a copper's wage. Seems the coppers there live a life of luxury, but it explains why he was eager to sort out the green card for the English mechanic so quickly. He tells me, 'If you get stopped anytime, tell the officers to contact me personally, but don't abuse it.'

I met his daughter, she was stunning, mouth-watering, and a real Miami Princess spoilt type, whom I ended up knobbing as she fancied the pants of a bit of rough, namely me. It was a one-off, and she took me out to dinner at this expensive seafood restaurant, which would have cost me half my week's wages.

Giancarlo wants to have a staff meeting at French's, a local Italian restaurant (yeah, called French's!) one Monday night. We all arrive sharp at 7.30 pm and are ushered to a table in the corner. The Guv arrives with two other BIG blokes, all broken noses and padded out suits. We all sit down and discuss how the business is going, when Giancarlo drops a bombshell, saying, 'Some motherfucker is stealing from me! I want answers now.'

We all look round at one another, and then he says, 'I have to make an example of this.' And with that Clint, one of my fellow mechanics, is picked up by these two big blokes and taken outside. What happened I don't know, but I never saw Clint again, and no one talked about it, ever. You read about the Mafia – was Giancarlo anything to do with that? I don't know, or want to know for that matter. All I do know is that everyone knew him, he was respected, and very well connected. Many people came to the garage, had appointments with him, and left with smiles.

Another thing I knew was that I never wanted to get on the wrong side of him.

Bill and I work our magic on some real old classics, including a Rolls-Royce Silver Ghost, immaculate and reputed to be worth a million dollars – no grease marks here, everything is covered up.

Life was pretty damn good. And did I mention the girls? In England they were so reserved, but in Miami they were really forward. I remember once going out to a bar with the boys from the garage after work and we get chatting to this girl. After a while she said, 'I've never fucked an English guy before.'

Well, not one to pass up a hint, I said, 'You won't be saying that tomorrow night.' I know, smooth! Talk about an East End boy putting it about, if there's a generation of short bald Floridians, now you know why!

The only downside was that I missed my family and friends badly. I talked to mum and dad weekly, but I was homesick and dad wasn't well, but I had to work. This goes on for twelve months and eventually I told Giancarlo that I needed to have a break, I needed to go home for a few weeks.

He looked into my eyes and said, 'You're not coming back, are you?'

I insisted I was, but he knew. The gent that he was, Giancarlo buys me a flight, one-way, and says keep in touch, you're always welcome here.

But I've never been back to Miami since.

So, back to the UK and no job. But what an adventure.

I spend the next few years doing consultancy work at various garages around North London, going back to what I know, but with me a new adventure is never far away. I was reading a car magazine one day when I saw an advert that took my fancy. It was for a company that had developed technology to allow engines to run on unleaded petrol without the need for conversion. Apparently it was based on technology originally employed in Spitfires in World War II and they were holding a seminar in the West End, so me being fascinated by engines and also being a bit nosey, I decided to go along.

The seminar was conducted by the three MDs of the company: an automobile engineer called Ted, along with Jonathan – who was an inventor and who had a family connection with the original Spitfire technology – and lastly Anthony. The event was effectively a sales pitch, a way of trying to recruit us as distributors of their technology.

There were about 150 of us in the hall, most of us mechanics, and of course I knew a few of the guys there. They went through all the history of this stuff and how the process worked by using lead replacements in the fuel.

It was basically a steel tube with an outlet and an inlet. The idea was, the fuel line was cut, and this device was attached to the cut ends, so that the fuel ran over the tin pellets inside it, which would supply lubricant for the valves, which would normally rely on lead contained in the petrol for this. They told us what their invention did, and how it increased the research octane number of the fuel and stopped it pinking (meaning 'de-detonating', which is abnormal combustion within the engine causing incorrect running) by supplying minute amounts of lead into the combustion chamber to stop the valves de-detonating. Normally, running a car whose engine was designed to use leaded fuel on unleaded fuel would of course cause pinking, the only cure being to have the engine 'converted', which was a relatively expensive procedure.

Of course it all sounded fantastic and would be an infinitely cheaper alternative to converting engines to running on un-leaded fuel, so could be a great earner for garages. This was the late eighties/early nineties, and about the time legislation was coming in about emissions and people were increasingly conscious about air pollution, and so converting engines from leaded to unleaded petrol was big business. Most engines require lead in order to run, you can feed them with unleaded for a time but soon the lead-starvation will cause pinking, and lack of performance and eventually seize up the engine.

The sales team were convinced it worked and that it had been tested at MIRA (the Motor Industry Research Association). At the end they invited any questions and of course I put my hand up and asked how much lead was released by these tin pellets. They said they didn't know. As they carried on talking I got more and more intrigued, because I could see the application and how it could work, but I wasn't a chemist so I didn't know what the chemical equation might be of what this tin/lead was doing in the fuel. The more I got into it the more questions I asked and the more I asked the less they realised they knew about their own technology.

We stopped for a break and while I was sipping on my tea and chatting to some guys I knew, Ted approached me and asked if we could chat afterwards. They told me they really appreciated my questions and that I obviously knew what I was talking about. So we got chatting and I told them about my credentials and the work I'd done on engines. Two and-a-half hours later they said they'd like me to become a consultant for them and do some testing, which I was happy to do.

This was on the Friday. On the Monday the engineer, Ted, turned up at the garage I was working at and gave me three of these units, each a different size. What I had to do was cut the fuel line on a car and insert one of these tubes (which looked like a long suppository), drive it for a hundred miles and then he told me the car would run on unleaded fuel without a problem.

So to test that the unit was actually working, and that the test vehicle did not just have an engine running on unleaded fuel – as it is capable of doing anyway for a while – I had to do an emissions test.

They claimed that this unit could reduce emissions by up to 40 per cent. The first engine I tested was in a Ford Cortina which runs on a '98 research' octane rating. If it runs on a lower octane rating, then you get pinking (also known as knocking).

So I did as Ted suggested with the Cortina, and after one hundred miles driving I switched over to unleaded fuel. Sure enough it worked as promised. Three hundred miles later I tested the emissions. The carbon monoxide levels had gone down, but the proportion of hydrocarbons (a component of unburnt fuel) had gone up, and there was a bad smell coming from the exhaust.

Something was wrong.

I got on the phone to the company, Effects Inc., and asked them some questions which, as per usual, they couldn't answer. So I told them about a good friend of mine who worked as a chemical engineer at Castrol, and asked if they would mind if he ran some tests on my behalf. This they were happy with, so I cut one of the units in half, emptied the tin pellets out and sent them away to my chemist friend.

About a week goes by, and my chemist mate phones me up. It turns out that inside the tin pellets there was lead, which was known to be more carcinogenic than hydrocarbons. In his opinion there was insufficient lead being absorbed into the fuel to lubricate the valves, but what lead there was in the pellets created more pollution in the exhaust gases than leaded petrol did. When I told Ted and the team about the findings they were horrified. It meant they had 3,000 units of stock built that were worthless and had to be scrapped.

They blamed the results on inferior pellets, so they got in touch with another company that could supply them with ones containing purer lead. Problem solved and I'm hero of the month, so they make me their international consultant. Effects Inc. has operations and connections in Canada, Cuba, Central America, places like that and they want me to go over to these places to give talks on what I found.

Meanwhile, the company is contacted by this Cypriot guy called Milton who is a world expert in CeO_2 – cerium oxide, which is basically a powder that comes from brick dust, but a hundred years ago or more this was used in gas mantles because it crackled and burnt with the fuel. What Milton discovered over several years of testing this material in car engines, was that by puffing small amounts of CeO_2 into the combustion chamber (the area above the cylinder in which the explosive reaction called combustion takes place) more fuel was burnt with no residue left, which increased power and reduced emissions: a win-win scenario.

When I heard about Milton's discovery it sounded very interesting, but the difficulty was getting the correct amount of CeO_2 into the combustion chamber with the intake. Too much and it nullifies it and it won't work, too little and you don't get the benefit, but Milton reckoned that these special jets he'd developed could deliver exactly the right amount.

However, when I tested the process I found these jets were all either too small or too large, and all the emissions results were going up and down. We were getting more chemical equations coming out of the exhaust than were going in! Also, the increased heat from the spark eventually wore out the spark plugs (these ignite the explosive gases in the combustion chamber).

So I contacted an engineering company in Portsmouth to make some needle jets in seven different sizes so that I could regulate the amount of CeO_2 going into the chamber. Now, cerium oxide is pretty unstable because if it's anywhere near dampness it turns to a mulch and the jets that we built needed a hole to suck in air. However with the cool air hitting the hot engine it created water droplets because of condensation, which clogged up the jets because of its combination with the CeO_2. So it was down to me again to invent a solution, which I did by drawing heat from the exhaust manifold and utilising this.

All the time I am getting paid a small amount by Effects Inc. as their consultant, but I feel that now, since I am inventing and providing them with solutions, I should be getting more money. We had a meeting and they agreed to increase my monthly fee. Happy days.

Milton now turns his attention to diesel engines because they're the greater polluters, but this means completely reinventing the units and that means even more work for me. After a couple of months, though, I notice that my money hasn't gone into the bank. I call them and they promise it's on its way, but meantime they need me out in Canada again. This I reluctantly agreed to and I went out there for three weeks.

A week into the trip my consultancy fee was due but didn't arrive, so I called Ted, who told me they had a few financial issues which had delayed my payment. I said while I was in Canada this was the only money I was earning and so I needed it. I gave him a piece of my mind, and he promised he'd sort something out. This was on the Friday and come Monday?

Sure enough, no money.

Tuesday? No money.

So I call Ted again and he says he'll sort it, and to call him back in an hour. So an hour later I call back and there's no reply. I try Jonathan, no reply either, same thing with Anthony.

Next I try their home numbers, mobiles, everything. Three hours later I am getting seriously pissed off. Eventually Jonathan answers his phone.

'We got a problem,' he tells me.

Oh shit, I think. 'What problem?'

'We've had to leave our offices.'

'So what's happening with my money?'

'Don't worry, we'll sort it out when you come back.'

'Fuck that! I'm coming back now.'

I packed my things and got the next flight I could, arriving back in London the next morning. I got straight into my car and went to the offices, which were indeed locked up. So I went to Ted's house and he opened the door, looking a bit startled. I told him that I wanted my money.

'Oh, but we can't,' Ted admitted, 'there's a problem with the company.'

'I don't want to fucking know,' I told him. 'Are these your car keys?' I asked, plucking the keys out of his hand. 'Right, I'm taking your car until I get paid.'

So I walked out of the house, bold as brass, got in his car and drove it round the corner, out of sight, locked it up and took the keys home with me. The next day I got my money in full, but the company was clearly in trouble and I couldn't be doing with all that shit every month, so I told them never to call me again.

Meanwhile, out in Canada, they had an associate called Brian who ran Canadian Effects. It was a different company to Effects Inc. so I was still happy to be in touch with him. He had discovered a fluid that was a surface modifier, which does exactly what it says on the tin: it modifies surfaces.

The principle of it is this: all oil, needed for reducing the friction between moving metal engine parts, breaks down under pressure and can no longer lubricate and cool. But by adding this special fluid it had the effect of increasing the level of pressure oil needed to come under before it broke down. Consequently this would help prevent engines from seizing up because of oil losing its efficacy, a kind of 'wonder drug' for oil.

However my recent experience had taught me that if something seemed too good to be true, it probably was. Oil generally degrades at a pressure of about 80 lbs per square foot, but sure enough with just a few drops of this 'snake oil' the resistance went up to 150 lbs per square foot, an incredible result.

So I asked Brian if I could have a sample of this magic 'surface modifier' fluid.

A week later it arrives in the post. It looks like runny honey, and has no smell whatsoever. Then I put a few drops into some standard oil and within seconds it is fully mixed with it. In fact it is a very rare thing to find a substance that will integrate with oil perfectly like that, so I'm even more intrigued.

Now, I'm a mechanic, not a chemist, but I know a man who is a chemist, so I send some of Brian's sample over to my good friend at Castrol for tests. I took the day off work and went up to Nottingham to see him, and the first thing he does is test it with the two best oils money could buy at that time: the Mobil 1 and Castrol GTX.

We test it and get to 160 lbs per square foot and it still hasn't broken down and my guy is mystified. He was sniffing it and looking at it, but obviously I couldn't let him do a full test of the ingredients because it was top secret, although he did tell me that one of the by-products of the fluids was obtainable from ICI, one of the biggest chemical companies in the world. So we knew that part of this blend was produced by ICI, but we didn't know what the rest of the ingredients were.

So this miracle substance is working but I need to look for side-effects. And the one thing we notice is that it's giving off fumes, even though the equipment lubricated by the oil isn't seizing up. So we test the fumes and discover that it's hydrogen chloride, which is dangerous. However, so long as it stayed within the gearbox or whatever mechanism it was added to, it would be OK.

But for me to put my name to this stuff, I needed to be absolutely sure it was 100 per cent safe to use, so I asked Brian to send me a Material Safety Data Sheet for the product, which would list many of the ingredients in this secret recipe. When I received it I forwarded it straight on to my chemist at Castrol, and he calls me back and tells me it contains chlorinated paraffin, and it must be this that is giving off the hydrogen chloride gas. Chlorinated paraffin is dangerous if it is 'short chain', but 'medium chain' or 'long chain' chlorinated paraffin is OK, so I checked with Brian and he assured me that his product is made with medium and long chain chlorinated paraffin.

All that seems reasonable but I want to do more practical tests. If it's having this effect on the oil, what is it doing to the oil seals in an engine, I wondered? I took two oil seals, one brand new and one old one which had hardened and split, and dropped them into this mixture of oil and the 'snake oil'.

Much to my amazement, it had no effect on the new seal,

which was great. Not only that, but it made the old hardened seal rubbery again – in other words it had brought it back to life. This really was miracle stuff!

I've seen all the additives and wonder drugs for cars over the years and they've nearly always been crap. My reasoning is this: if you've got a multi-billion pound company like Castrol with teams of developers all over the world and they haven't managed to find the secret recipe yet, what chance has some small backyard outfit got to do so? But this guy Brian who worked for Canadian Effects... well, he seemed to have done it.

Next I tried the fluid in gearboxes that were whining, and engines that were noisy, and it did the business. I tried it with transmission oil and it integrated immediately with that too. Every test I did it passed.

Brian even sent me the results of tests that other companies had done, including big shipping companies over in Canada. It just so happened that, after Brian had sent me paperwork from one of the companies, I went over to visit a few months later to do some applications for them on their heavy machinery, and in the paperwork I gave them were copies of the test results they'd carried out. They called me up and said that these test results weren't correct; the tests they'd carried out showed no improvement of performance at all.

This made me suspicious. But nevertheless every test I did, Brian's fluid passed with flying colours. This was until one day I got a call from a UK company who had been using this stuff and an engine had virtually seized up, so I went down to Dorset to take a look. They said that the engine had been running noisily and been burning off a bit of oil. So they thought they'd try this 'snake oil' they'd heard about, but the engine had seized almost immediately.

Strangely, this was completely the opposite experience to every test I'd done, so I was baffled. I called Brian and he said he'd get back to me, but he never did, so I decided to do my own investigations. I asked the company in Dorset if I could take away the fluid they'd got, so that I could compare it with the original samples I'd been sent. I noticed that it was a slightly different colour: the old stuff was lighter than the new bottles they were selling. Once again I sent it off to my man at Castrol and he came back to me to say that the new sample contained the dangerous 'short chain' chlorinated paraffin – not the medium or long chain chlorinated paraffin that Brian had told me they were using.

I called Brian and presented my evidence to him and he eventually admitted that to try to reduce the cost of producing this 'snake oil' he had changed from medium and long chain chlorinated paraffin to short chain. The result wasn't just dangerous; it also completely reduced the effectiveness of the fluid. I felt terrible because I had been duped and, worse, I had encouraged other businesses to spend a lot of money on this stuff.

There was one other issue that we had with this snake oil, though, and that wasn't Brian's fault.

Brian gets a call saying that an engine-driven power station in Central America keeps seizing, so I get sent over to take a look. The trip is a nightmare: London Heathrow to Houston Texas, then a wait for four hours to get another plane to El Salvador. This was a total trip of fifteen hours and by the time I get there I'm fucked. It's now 10 pm at night, still blazing hot outside, and I'm sweaty, tired, jet lagged, and ready to kick arse.

I'm at the airport, and I can't see the guy – Christian Auer – who is meant to be meeting me. I have never met him, speak no Spanish, my mobile does not work and I have no local currency to call him from a call box. I ask at the airline desk, tell them my woe, and they call him for me. At last, I'm connected to him.

'WHAT?' he says. 'You're not due till tomorrow. Stay there, I'm on my way.'

Thank FUCK for that, I think, as I'm getting dodgy looks from the locals, who are eyeing my suitcase etc. So now it's out to customs, where I meet a customs man who speaks in broken English.

'Please open your suitcase,' he asks me.

Inside I've got new shirts, short sleeved ones, trousers etc., all brand new for the trip – they've still got the labels on. This guy goes through them, then says, 'Are these for your personal use?'

'Yes,' I reply.

'Well Señor, you could be bringing them in for resale, so you will have to pay $400 to go through customs.'

NO FUCKING WAY JOSE, I think to myself, but he is adamant. I have to pay or he won't let me pass through customs. I explain where I'm going, what I'm doing, and mention that the father-in-law of Chris, this guy I'm meeting, is the Minister of the Interior, Don Roberto Machon.

After that he salutes me and says, 'Have a good trip, Señor.'

Amazing what a bit of name-dropping does, got ya, fella?

Chris finally arrives one hour later, apologises, and we are off to my hotel, a little secure guarded eight-room place called Casa de Rada. It's beautiful, all flowers and plants and a lovely lady to greet me called Stella, who is the owner.

I unpack, shower and now it's gone midnight, but Chris wants to take me to dinner. I would love to just go to sleep, but I'm Hank Marvin (starving) so we go to a small fish restaurant where he orders something for me off the menu. It's called Concha de Burro and it arrives in a tall glass, and looks like mussels. OK, I think, I can handle that. Until he squeezes lime juice over it and, oh my God, these little fuckers start moving!!!

'What the hell is this?' I ask.

'Just taste it Bernie, I promise you will love it.'

Well here goes, and fuck me it's great, all fishy and slimy, but so good. We have a beer and fresh bread, and veggies and I'm full, tired and want me bed. At the hotel, I shower again, it's still really hot, and I'm informed that the daytime temperature is about 46 degrees, and at night it's about 38 to 40 degrees. I try to sleep, but my stomach is going ten to the dozen, I've got the farts, rumbles, and I wanna be sick. I just make it to the shower when my guts repel all I have eaten, all over the wall and floor. UGHHHHHH!!!

That's better, I think, except for the clearing up, which I do. Then I sleep till 7 am, shower, go down for breakfast, just toast and tea, and then Chris arrives. We take a three-hour drive in a clapped out Peugeot, that's got a rail from floor to ceiling with shotguns attached, and we're in the middle of the jungle, at a place called Ahuachapan (near the border with Guatemala), near the power station we've come to fix.

On the way Chris tells me his story. He was born French-Canadian, ex-French Foreign Legion and ex-mercenary, settled in El Salvador some twenty years ago, met a lady, married and had kids, and the only other country he can go to is Canada. Fuck me, I think, sounds like he's a hired killer for the government. I'd better behave myself!

In the heat, we gradually check over all the engine parts of the power station to see what caused it to fail. The bearings are virtually blue, and that's because of lack of lubrication, but I could see that the oil that was drained out was plentiful, so it wasn't oil starvation that caused it. I'm working with five engineers, who speak no English, so I talk through Chris, who has to translate. We eventually get the engine back in place and up and running.

But after two days, it seizes again.

WHAT THE FUCK IS GOING ON??? I wonder.

I go back to the plant, and I don't understand this at all. We checked everything out and all I could find was that the number 1 and number 12 big ends (the bearing surface between the larger end of a connecting rod and the crankpin of the crankshaft) had seized up, which was due to blocked oilways (the channels that permit the flow of lubricating oil).

Now I had a suspicion, and I want to do an oil analysis, so I take a one-litre sample of it and Chris and I drive to Guatemala City, which is about four hours away. We stop at one of the laboratories and they say it will cost 300 quetzal (the local currency) and it will take two days for the results. But for 1,000 quetzal we could get it in four hours. We take the second option and after two hours we return, and are told with a smile, 'Señor, this is piss.'

I say, 'WHAT?! Is this fucker swearing at me?'

'No, Señor,' he assures me. 'I speak English good. Your engine oil has 35 per cent human urine.'

Now the oil being used in this engine costs over £250 to fill it. It turned out that what these engineers at the plant were doing was filling the base with oil and then peeing in the intake to bring up the level, then adding cheap reclaimed oil, and then selling the new oil! Cheeky fuckers.

We take our results to the plant manager, and, after a further two days of building the engine, all the engineers are sacked, and the correct oil is fitted. To date, there has never been another problem reported. They're happy, and I get an intro to a company in India.

My reputation is spreading. I do more work and am flown to Bombay, Delhi, Karachi, Calcutta, to do similar work on outside power plant engines. So I land in the UK after being away for two months, tired, happy and dying for an English cuppa and a kosher bacon sandwich. I've got more air miles than Alan Whicker, but I get yet *another* call to go to El Salvador. This, you'll remember, became that infamous trip I mentioned earlier, when there was a military coup, and when I lost my mum.

After all this I was done with travelling for a while, but I knew I'd been lucky to experience so many other cultures. In El Salvador, for instance, the law of the land is that everyone carries a gun or machete and you protect your own property. If someone is stealing from you, then you shoot to kill, not maim.

Chris, as I've said, was an ex-mercenary, so he knew what he was doing in these matters, and he always carried a gun strapped to his ankle as well as one in his jacket, and he kept a sawn-off shotgun in the car. He was married to the daughter of the Minister of the Interior. One day she was in the car with their two young children when a guy ran up to them, opened the passenger door and tried to steal her handbag, but before he could grab it she shot him straight between the eyes, right there in broad daylight.

Obviously, Chris's wife didn't know if he was going for the handbag, her or the kids, and she wasn't waiting to find out. The culture over there is 'another dead Indian, more tortillas for the rest of us' and with her father also being the Minister of the Interior, no more was ever said about it. In countries like that death is part of life, so to speak. It was like the Wild West sometimes.

By contrast, Cuba couldn't have been more hospitable and, of course, you don't need me to tell you that with all those classic cars driving around I felt like I was in heaven! There were 1959 Ford Zodiacs that looked like they'd just come out of the factory, and everyone was so proud of their cars that they kept them in perfect condition, no matter how old they were. The people out there didn't have much, but they were rich in generosity and friendliness. They were under tremendous pressure all the time to eke out a living, but they were genuine and worked hard.

I am sure that Cuba can be just as dangerous as El Salvador at times, and there is plenty of poverty, but I never received anything but a warm welcome. I was there to help the local distributor, Gary, promote the Effects product, but he would often find himself in trouble with his repayments to the company, so Effects were continually threatening to go over and seize his stock. He rightly pointed out that if they wanted to come over to Cuba and take stock back out of the country, he'd love to see them try!

Gary was paying the company what he could for this 'snake oil' and so were his clients. Sometimes they were paying money they couldn't really spare, and it was those guys, the customers and Gary, that I felt most sorry for when I realised I'd been duped. It was them who were going to be left with broken engines on which they relied for their livelihoods. I regret it to this day.

With my foreign adventures over for the time being, over the next few years I worked at various garages around London as a senior mechanic, foreman or consultant, but it was always the same: when work dried up the most expensive employee had to go and that was usually me. So eventually I took the bull by the horns and decided to open my own garage.

Above left: My mum, God bless her, when she was 23.

Above right: Mum and Dad on my wedding day.

Below left: With my daughter Lisa Marie in 1983.

Below right: Lisa and me with baby Nadine in 1987.

Above left: I'm surrounded by women! With my girls in the lounge in the early 1990s.

Above right: Nadine and Jamie on their wedding day with Jo and Des, the best parents-in-law any man could wish for.

Below: Leepu and me in the 'Angel car', a Bangladesh wedding car we created.

Above left: With wonderful bloke Jools Holland. Leepu and I made his dream car, a Jet 1.

Above right: 'Angry Frog' car based on a converted Mitsubishi that runs on used curry oil.

Below: The *Classic Car Rescue* team (left to right) Big Diago, Sat, Mario, me and director Steve Scott.

Left: The Corvette Stingray restored by the *Classic Car Rescue* team.

Right: The team with the most expensive rescue job: a Ferrari Mondial.

Left: Do you really wanna get me annoyed?!?

Right: Lisa and I with our best friends in the whole world, Sharon and Phil Wilson.

After a few weeks searching I found a small garage in Belsize Park. So here I am, I have my own client list, a couple of cars ready to go, and bang, crash, wallop, I'm in business. I've got two ramps there, a small space out front for parking, a little office at the back and flats above, so there's no working after 6 pm, as I don't want to disturb the new neighbours. I decorate the place all bleach white, red ramps and half the walls red. I call up some clients and away we go.

First thing Monday morning two cars turn up, both Jaguars, needing a service and some other stuff. I call the local suppliers and yes, they will supply me on a weekly basis and with a limit of £250-worth of goods.

First two jobs done, I'm paid and have a tidy profit of £350 – not a bad first day's work. Unfortunately, during the rest of the week only two other cars come in. Suddenly I remember why I never went into business before. I'm shitting it, my rent is £1,000 per month, plus electric, ground rate, fuel, etc., so I have to take at least £1,500 to break even.

I start making desperate calls to all the clients I have worked with over the past few years, morning and night, and slowly the work comes in, and I mean really comes in. Within a month I am stacked, three or four cars a day, working on my own, six days a week. I can't cope.

Just at the right time a guy comes into the garage. His name's David and he is from Australia, says he is a mechanic with ten years' experience and seeking a job. We chat, he seems a nice guy, so I take him on: one month's trial for £250 per week. Things go well for the month, we get on well and he is a good mechanic, turns up on time, no days off and always helpful. I tell him, 'OK, £350 per week, that's the best I can do for now.' We shake hands and I even give him a set of keys.

After a month or so he asks if can he do his own cars on

a Sunday, and I say no problem, just don't make any noise or the neighbours will complain. I remembered that I was always grateful to the garages I worked for in the early days for letting me do my own work on the side, so in the same spirit of generosity, I had no problem with David doing work for his mates in his own time. Four months down the line, David is doing well. He does his mates' cars on Sundays, always cleans up and leaves by early evening.

Everything's running along fine until one Sunday I get a call from a client who says he's just driven past my garage and it's stacked to capacity, and there are cars queuing down the street.

What the fuck?

I jump straight in my car and head over. David and three other guys have cars jacked up in the street, both ramps working. All the lights are on and there's loud banging, engines revving and I notice two of my clients' cars are being worked on. I park around the corner and walk in.

'What the fuck is all this about?' I ask him.

Red-faced, he tells me it's his private work and the other mechanics are mates who've come to help him. I ask him what my clients' cars are doing there on a Sunday, so he says they asked him to do them privately.

'REALLY???' I say. 'We'll see about that.'

So I call them up and find out that this little fucker has been calling my clients, telling them that if they bring the cars in on Sundays it will be 50 per cent cheaper. I then call my supplier and I find out that all the parts he is using have been added to MY account. He had a right little business going on: no overheads, a ready list of clients and when they paid for parts he was putting it on my account and pocketing the cash himself.

This is not on, so in the politest way I can muster – which isn't very – I tell him FUCK OFF now, take your scrawny mates with you and don't come back. I even take his car keys, as I'm keeping his car until I find out how much this cunt has had me over for. He then calls the police, who attend the garage twenty minutes later. I tell them the story, but they insist I give him his keys back, which I do. They escort him and said mates off the premises, never to be seen again.

With bills mounting, and all the parts he has nicked off me for his private work, I'm up shit creek without a paddle. I have to find some £4,000 just to get on track with suppliers and rent etc., so – fingers to the grindstone – I work and work all on my own, night and day, until I'm straight again.

More than anything else, this episode made me realise that I'm a good mechanic, but no businessman, so after a year I decide enough is enough. I close the business and go back to working for someone else – let them have the headaches of dealing with staff and paying bills. If you'll worry about the money, just give me a team of mechanics and I'll make any garage run properly.

CHAPTER NINE

BERNIE'S NO-NONSENSE GUIDE TO BUYING A USED CAR

Apart from buying a house, buying a car is often the biggest purchase you ever make. Even a second-hand car can cost thousands of pounds, so going to buy a car is an exciting thing, your blood is pumping, the adrenaline is rushing, because you know you're going to part with a lot of money and the stakes are high.

The thing that most people do is that they buy on impulse, which is wrong. It's love at first sight: you see the car and that's it, it's an orgasmic feeling: 'I've gotta have that car!'

But let me tell you, all that glitters ain't gold. A shiny car in damp weather can hide a multitude of sins.

Rule number one – never buy a car in the rain. When a car is wet it looks very, very shiny and you can't really see if there are a lot of ripples in the body and scratches are hard to make out too. Without the rain disguising it you can see if the paintwork is completely flat – and if that's the case, then it hasn't been looked after, and you could be in for a re-spray.

So never buy a car in the rain: besides getting your hair wet, you could also be buying a pup.

Unless a car is made of fibreglass, when you go and see it for the first time, keep your distance from the salesman and take a magnet with you. Assuming the car's made of metal, go over each panel with a magnet – if it doesn't stick to the doors or wings then there's a good chance it's got a lot of filler (glass fibre body repair material) in it, meaning it's probably had an accident.

The majority of cars today are still made of steel. They may be a lot, lot thinner than they were years ago, but most cars on the road are still metal. However, some makes of car use various alloys and some are carbon-fibre, so if you're not sure about a certain vehicle go online and look up the spec – it will tell you what composite materials the car is made from. If it's a metal, then the magnet should stick.

Once you're happy the car is mostly made of what it's meant to be made of, the next thing to do is ascertain whether it is straight. Crouch down by the rear light and look down the length of the car. If it has lots of ripples in it, it's been badly repaired. If it ain't straight, or the doors aren't aligned with the panels, you're in serious trouble.

How much trouble? Well, hopefully not this much. Many years ago, we're talking probably early seventies, the gorgeous and wonderful Barbara Windsor bought a Rolls-Royce for her old man (at the time) Ronnie Knight. A couple of months later she brings it into the garage, and says she's having to change the tyres every three weeks! They were wearing out as quickly as they could put new ones on.

When I checked the car over, the first thing I did was look straight down the wing and saw that the chassis was as bent as a nine-bob note. Not only that, but the steering geometry and components had been straightened, not replaced, and the thing was driving down the road crabwise. Bloody dangerous! You didn't need to be a mechanic to see this thing had been in a serious accident and badly repaired, so Barbara took it straight back to the guy she bought it from, but – surprise surprise – he didn't want to know.

Next day I went down there and read him the riot act, and, Bob's your uncle, he reluctantly gives her the money back. She'd bought it on impulse, all shiny and new so she thought, but obviously she hadn't looked at it straight on. As for the dodgy dealer, well, all I can say is he should think himself lucky I went down there. There were one or two people whom Barbara knew in the East End at that time that could've done a lot worse damage to him than I could.

Another tip for when you're buying: put your hand on the bonnet. If it's warm, then obviously it's been recently driven. That could be genuine, or it could mean that in cold weather it doesn't start easily. That could mean anything from a fuel injection problem to a low compression problem within an engine. Some cars smoke more from the exhaust when they are cold than when they are hot. Sometimes people try to dupe you and have the car warm when you get there, so be aware of this, and don't be afraid to say you'll come back when the engine is cold.

They say seeing is believing, and sometimes you can't believe what you see. If a car is too good to be true then it probably is. Do a bit of detective work when you're looking round a car, think about its age and mileage and think about what sort of state you expect that car to be in. A little bit of wear and tear isn't too much of a problem if it's an older car, but wear and tear on a low mileage car rings alarm bells for me.

For instance, if a car has 20,000 miles on the clock, have

a look at the brake and clutch pedals. If the rubbers on them are very worn away and smooth, then beware. Unless you've got size 25 shoes I can't see how the rubbers could wear away that quickly. Look at the steering wheel. If it's worn away or pitted then again it suggests the mileage on the car isn't exactly what it's showing. Look at the general condition of the seats; if the 'squab' or the back part is worn away then you've either had a fat bastard driving it or the mileage isn't what it should be.

Listen to the engine. It should be basically quiet. If you can hear strange rattles or squeaks, make sure you find out what the problem is. They might just say it's a squeaky fan-belt, but it could be your water pump on the way out or crankshaft damage, which could be a very expensive repair.

When you take it for a test drive, look out for certain things. If it's a manual car make sure the bite on the clutch isn't too high. When you put it into gear and let the handbrake off and lift the clutch off to make the car move, the bite should be an inch-and-a-half or two inches from the bottom when the pedal is fully depressed. If it's right near the top then that's telling you the clutch is on the way out. On some cars this will cost you £150 to replace, but on others it can run into thousands, particularly if it is a single mass flywheel or if it's a dual mass flywheel.

Go through all the gears, take it out onto a national speed limit road so you can get up into fifth gear or sixth gear if it has one. A rumbling sound in certain gears or a car jumping out of gear can denote a very, very serious problem inside the gearbox. For an automatic car, put it into gear and drive, and then watch the change-up speeds: first to second should be roughly at 15–20 miles an hour, and it should be into top gear at 40. If you're getting faster and faster and you're still going through the gears, in an automatic, that's telling you the clutches inside are worn. Again, a very expensive repair.

Also, and this is the thing everyone forgets but I've seen it happen before, make sure the car goes into reverse! I had a client eight or nine years ago who bought on impulse. It was a really beautiful car but it had no reverse gear, so it was next to bloody useless! It was a Jaguar and it cost him two-and-a-half grand to get it fixed.

Look under the bonnet. If it looks too clean always query it, ask if it has been steam cleaned. Sometimes it can be done for a genuine reason and the seller wants to present something cleanly, but sometimes engines are steam cleaned in order to disguise things, particularly major oil leaks. Engines that are steam cleaned and finished with a bit of WD40 (an oil that acts as a water displacing spray) may look nice, even smell nice, but a few miles down the road you may find you've got a big puddle of oil underneath it.

Not everyone carries car jacks with them, so obviously to look under the car for most people is very, very difficult. My advice is, stick to what you know: if you're in computers don't do someone's haircut! Get friendly with your local mechanic, even if you're buying from a dealer, and take him or her along with you. Offer to pay them for their time, even if it's just a nice cup of tea and bacon sarnie afterwards, but get him to check out what you can't check out.

I'd say that the majority of the people in the motor trade can be magicians when they want to be. They can hide anything that you're not going to find in a million years. So take someone with you who knows what they're doing and that you can trust, and make sure they're happy with the car. But even if they are, always do an HPI check; that way you know you've done all you can and you can be confident you're not going to get, what we call in the motor trade, 'FUA' – meaning 'fucked up the arse'!

All cars that have been in an accident have to register with HPI. Go online, enter the registration number and it will give you a detailed history of the vehicle and if it's been in an accident. If it's been in a bad accident then walk away, because even if you can buy it at a reasonable price, you'll probably not be able to re-sell it.

It's very important to make sure the car has a service history. If you're buying a car you want to know that whoever has had it before has looked after it. If it's not got a service history, query it, and find out why. If the dealer hasn't got it, ask them to contact the previous owner and find out if they have one. If a dealership has been carrying out the work then they will hold a duplicate detailing all the work that has been done. Again, you'll struggle to sell a car without a service history, so think twice when you're buying it without one.

Ask how many owners it's had. If you're buying a 2010 car and it's had five owners already, when you come to sell it as the sixth owner it ain't worth nothing.

Buying a car from a dealership, to my way of thinking, is safer than buying from a private person. An individual selling a car can be one of two types of people: they could be genuine, someone who's selling his own car or doing so for a family member, or you could be dealing with what's known as a 'home trader'. A lot of home traders don't register themselves as traders, they buy from auctions and sell to the public and if anything goes wrong with the car you've got no comeback whatsoever.

If you buy from a dealer then you are covered for a statutory warranty of three months, and that's by law. They may well offer you an extended warranty, which will cover the main parts of the car, that is the engine, transmission and differential, but beware. If you take the extended warranty then this often renders the statutory warranty null and void, so anything else that goes wrong that isn't covered means you'll find it exceedingly difficult to get your money back.

Even if it's a Ford dealership and they sell you a Jaguar it makes no difference, they are a dealer and you are covered by your three-month warranty. Anything that goes wrong in those three months, apart from consumables obviously like brake pads, and you are covered by law: either get your money back or they have to put that vehicle right. My advice is don't take the dealer's extended warranty; get your statutory three months, then when that's nearly up, shop around. There are loads of warranty companies on the internet and you can probably find yourself a better deal without infringing your statutory rights. That way you get two bites of the pie rather than one, and I'm a greedy fella!

If you have a car for sale it is better to sell it first and become a cash buyer. That way, you're in a much better bargaining position and likely to get a better price on your next car.

And if you're buying from an individual, ask them if it's their car and if it is registered in their name. A good trick to find out whether they are a home trader or not is when you phone up, say you're 'calling about the car they have for sale'. If they say 'which one?', put the phone down, as they are probably buying and selling as a business.

If you're buying from a dealer, ask if it has had a 42-point check and are they going to service it before you drive it away. And always ask if there's a deal to be done – nine times out of ten there will be. The best way to wangle a deal with a dealer is to be straight with them. Say, 'I'm interested in the car, the price doesn't interest me, I haven't got a part exchange. If I pay cash for the car what sort of deal can you do?' If they say, 'This is my final price,' say 'Fair enough, I'll spend my money somewhere else,' and walk away. They'll either change their mind and say something can be done, or they've genuinely got a very small margin on the car and so can't budge on price.

Be clear with a dealer by saying you might not buy the car today. What they wanna do is wrap you up, tie you up and get the deal done. But don't be strong-armed into a deal and don't buy on impulse, take twenty-four hours to think about it. Always keep yourself a little bit back, don't be too obvious, don't be too eager. If you seem that way then you've got the word MUG written on your forehead and every dealer loves a mug.

Beware of what we call 'dealer spiel'. They'll butter you up, give you compliments, saying you look great in this car, this that and the other, so you've got to be straight with them to show them you're in the box seat. Say you want to take it for a road test and on a fast run, so that you can feel how the car handles. Tell them you're not going to buy the car today and you're not going to pay the price they are asking. That way they know you know what you want and their sweet-talking won't work. They don't hypnotise you, they just use a bit of psychology, they butter you up in order to get a quick sale, and the quicker they can wrap up a deal the happier they are, and the quicker they can buy other cars.

Always get an insurance quote before you buy a car too, so you don't get any nasty surprises. It might only be a 1.6 litre you're looking at, but if it's a turbo or fuel injection this can have a huge bearing on the cost of insurance, so that bargain car may cost you more than you bargained for. Find out what road tax group it is: there's no point buying a cheap 4x4 if you're going to get lumbered with a £600 road tax bill and a £2,000 insurance premium. A more expensive car might be cheaper in the long run, so work out what your budget can do.

When you are buying a used car, it's second hand, it's been driven, so don't expect a ten or fifteen-year-old car to be in mint condition. Perfect cars have either been completely rebuilt after an accident or wrapped in cotton wool and not driven. If you asked me is it better to buy a high mileage car or a low mileage car, it's six of one and half a dozen of the other. A car that's been driven in town only, that's five years old and has 8,000 miles on the clock, is actually more worn than a car that's spent all its life on the motorway and done 50,000 miles. I'd rather have the one that's lived on the motorway – all that stop-go stop-go in the other one creates wear and tear. Plus, if it's got 50,000 miles and been serviced regularly you've got all the little quirks out of the way.

Buying a very sporty 'muscle' car is like having a beautiful woman, and I guarantee you she's been used more than an ugly woman! If you're buying a sports car, expect it to have had a reasonably hard life, they're there to be driven, they have power, so obviously someone has used that power. I'd say about 80 per cent of the muscle cars I've worked on – Ferraris, Maseratis, Lamborghinis – have all had an accident. They've all been driven fast and the people who buy them first-hand don't know how to handle them.

It's the same with anyone who buys a sports car. There's no point getting out of a Peugeot 206 and jumping straight into a Ferrari, you'll crash that thing on the first day. It's lovely to say to someone, 'I've got a Ferrari,' but I'd answer, 'Yeah, but can you drive it?' If you're going from a standard car straight to a muscle car, the best thing you can do is take some lessons with the Institute of Advanced Motorists, so you learn to control that thing. It's like going from a 100cc motorbike to a 750 twin motorcycle: if you're not careful and experienced you could end up killing yourself and other people.

And it ain't just muscle cars you need to be wary of. You should get specialist advice on any car you buy that is different from what you're used to. A few years back the wife of a customer of mine treated herself to a Mercedes SLK Automatic – nice car.

But she'd only ever driven manual Fords and Vauxhalls before, which were nice and easy. She bought the car on impulse (don't they always?), had no one with her, and then proceeded to set off down the M1 back from Northampton to north-west London, having no understanding of how an automatic vehicle works. After eight miles down the motorway with the lever set to a low gear, car screaming blue bloody murder, and clouds of smoke pouring out from under the bonnet, she pulls into the services and calls me. I send a truck straight up there to pick her and the car up and bring her back to the garage. The damage? Head gasket blown, automatic transmission overheated, a right mess. She got a £4,500 bill for her trouble and no, the dealer wasn't liable, it was just a very expensive mistake on her part. She learnt her lesson the hard way. Always look into things you haven't driven before.

Finally, once you've done the deal, check what you're getting. Make sure you have the logbook and all the keys. There should be two sets of keys. If there's only one then alarm bells should ring.

An old scam that still happens is someone buys a car and takes it home. Next morning they wake to find the car has vanished. They call the police, get a crime reference number and inform their insurance company. Wondering if the seller has something to do with it the police turn up at the vendor's property but they're nowhere to be found, the house is either empty or squatters are living there. They've kept the spare set of keys, know your address from when you filled out the log book retainer, and have driven off with it in the night. By the time you've noticed it's gone it's probably on a ferry to Calais. So make sure you have both keys when you purchase the car, as all modern cars have electronic management systems, and you have to show evidence to a dealership (log book, passport and utility bill) for spare keys, which are coded to the car.

I've made millions of mistakes in my life, but when it comes to buying a second-hand car I'm very, very thorough, as there's just no point taking risks. In fact half the dealerships in London won't deal with me, and if they do it's, 'Bernie, take the car away, tell me what needs doing and deduct it from the price, but for God's sake don't bring the bloody thing back again!'

They know if there's a problem with the car I will find it, and they don't even argue with me because they know that they're selling cars every day with minor faults that aren't picked up. The problem is people buy on impulse and they don't want to see the faults on a car, they don't wanna know. So if you come to me with a problematic car you've only bought a few weeks ago don't expect any sympathy from me, you've obviously not done your homework!

But follow my guidelines and tips and you'll give yourself the best chance of getting a good deal and a reliable, safe car.

CHAPTER TEN

BERNIE'S GUIDE TO CAR SCAMS

I am ashamed to say that in this day and age there are so many garages out there that aren't kosher. I'd say about one in four is probably not as honest as it should be. This is because either money is so tight that they've got to screw people over to make a decent living or they're just run by just the type of person who is greedy and doesn't care about screwing people over.

In north-west London for every twenty garages I could show you eighteen that would fuck you up the arse before you've even left the yard and they'd laugh at your back, they don't give a fuck. If you try to come back and complain they've got dogs and they'll see you off the premises. There's no licensing, you see. So, if you've just moved to an area, how do you know what's a good garage?

First thing to do is make friends with people. Ask around, find out which garages are popular, who do people recommend, are there any positive stories or bad experiences? Just check before you ask that they have no association with the garage. It's very important when you go into an independent garage to ask if they work to the guidelines of Trading Standards. That way you are protected. Also ask if they are members of any professional bodies like the Retail Motor Industry Federation (RMIF), the Institute of the Motor Industry (IMI) etc.

Should you have a bad time with a garage – for instance, if they don't do something properly and it needs to be put right – then you have some recourse. You'll have to go through local Trading Standards but the RMIF and IMI have client helplines and they will act as arbitrator between you and the garage to put things right. If they're not members of any association and don't work to the guidelines of Trading Standards then walk away. If you're not sure, then phone your local Trading Standards and ask which garages they have listed as working to their guidelines.

A nice smile and a cup of coffee in a nice palatial reception area doesn't necessarily mean it's a good garage or that your car is going to be looked after. Always ask questions, and if something doesn't sound right then don't be afraid to say, 'No thanks, please put my car back together,' and take it away.

Don't be duped by sponsored schemes where members are just those that choose to stock particular products. There are lots of membership schemes like this: you pay your money and you go on to their 'trusted' list – they don't check to see if you're any good or not. Blow it out your arse.

So you're driving down the road, thinking your car hasn't been serviced for a while, and you see a sign saying 'FREE tyre check, FREE clutch check, FREE brake check'. Well take it from someone who knows, there ain't no such thing in life as a free meal. It's just a way of getting you in, they'll take the wheels of the car and they'll tell you that you need new discs, new pads and whatever. If you decline to have it done you should not have to pay one penny, but don't be surprised if you get an earful, first telling you how dangerous your car is, then they'll get really angry when you walk out the door. If they do then you know they're not an honest garage.

An honest garage always puts the customer first and accepts it's your decision. But these places, where they try to force you to part with money, are often only paid pennies; these poor buggers rely on commission from sales to make a decent living. But to my mind that's no excuse – so beware. Don't be hassled into something you're not sure about.

One of the most common scams, and this often occurs in 'fast fit' sort of places, where they have very high turnover and try to get you in and out the door as quickly as they can, offering you cheap rates on basic services, is when they say they'll do you a cheap oil change. And there's a reason why it's so cheap – they don't bloody do it! They might change the oil but not change the filter, or perhaps put some peroxide in with the oil to make it look lighter and cleaner, so it looks like new oil when it's the old. That's lazy stupidity, and in my way of thinking it's fraud.

I have very personal experience of these 'fast fit' places. I went away to work for two weeks and one evening I get a call from 'er indoors. She has a real go at me, says I'm trying to kill her father!

You wot?!

Before I went away I serviced his Audi, put new tyres on, shocks, brake pads, discs, the works, then did the MOT. That car was spotless. While I was away he got a puncture and went to one of the 'fast fit' places in London. Whilst there he was told that the car was dangerous to drive, and that it required new front brake discs and pads, two tyres and had front shock absorbers which were badly leaking.

My father-in-law is a trusting sort of chap (others might call him gullible) and of course he said please do it. Four hours later, and a bill for £750.20 inc. VAT, he drove away, told my wife and she went spare. On my return I went to see the 'fast fit' place. I was recognised by the manager who had seen my TV programmes, I told him my woe, and red-faced he admitted that they had been rather over-the-top with the repairs.

No shit, Sherlock!

When I told him I had serviced the car the week before, I got an apology and a full refund. Seems they rip everyone off to get commission on the service and repairs they do.

Of course, not everyone can be recognised as a mechanic off the telly, so if you are suspicious about the work being done at a place, before you take the car in, just mark the filter with a little bit of nail polish. If the dab of nail polish is still there when the car comes back then you know it's still the same filter and not been changed. If they can't even be arsed to do this, the most basic of jobs, just consider: What else have they not bothered to do?

Always ask what's included in the price of a service and what sort of service it is. A full service should include changing the engine oil, and new oil filter and air filter. If it's a petrol car then the spark plugs need changing as well. Don't be blinded by seemingly great deals, make sure they are going to do the work you'd expect – if it's a 'full service' make sure it really *is* a full service, including a diagnostics check. Most new cars have an on-board computer which can scan for problems, so make sure this check is included in the price, otherwise you might find an extra £50 added to your bill you weren't expecting.

Brakes are the things that most garages will try to scam you with because it plays to your fears. You think brakes are the most important part of a car, which they are, so you don't want to mess about with them: if they need changing you think you'd better get them changed. But before you do, check back in your records. Ask the mechanic what the thickness of the brake pads is, and get a quote for the job. Then phone another garage and ask them whether they'd expect the pads will have worn down that much in the time since they were fitted and get a quote from them. Always shop around and get a second opinion where you can.

If a part needs replacing, always ask for the old part back. Sometimes the part might be perfectly useable and they'll charge you for putting a new one in, then put your old one in the next poor mug's car and charge them for a new part. And before a new part is fitted, always ask if it's a genuine part or a pattern part. A pattern part is made by a third party manufacturer, not by the manufacturer of the car itself, and although cheaper, can be of inferior quality. It can be cheaper in the short term but more expensive in the long run if it doesn't last as long.

Thank goodness, there aren't many dodgy garages about. By that I mean garages that will purposely sabotage a vehicle in order to take money off you, rather than just doing the odd sharp practice. But a favourite scam for those that are dodgy is urinating into your radiator. Maybe you didn't want to buy anything, so they thought they'd teach you a lesson, or just did it to ensure you come back soon to actually get some work done. So you get in your car and drive away, but within a few minutes you realise something isn't right. The smell is disgusting, and it takes some time to flush that out of the cooling system.

At another garage I saw a mechanic emptying valve grinding

paste into the oil. Now, that will act as an abrasive and wear the engine away and there's not much you can do after that but replace the whole engine.

Another nasty one is to put a sponge in your fuel tank. Years ago they'd use sugar. So the sponge is in there, the pump activates, traps the sponge and stops the fuel from coming through. You're driving down the road and all of a sudden the engine will cut out. You'll turn it off, the sponge comes away, and allows fuel through again, so when you turn the engine on it'll run for a short distance before the fuel is blocked once again and you're stumped.

I've seen cars come in with a misfire and the garage has said the head gasket has gone, so they give you a big quote for all these new parts and get your car up on the ramp. All they're really doing is steam-cleaning the top end of the engine to make it look like it's been taken apart, but the misfire is in fact just a plug lead or one of the coil packs has gone down. But they have charged you for a much bigger job. It's fraud, whatever way you look at it, it's fraud.

A rogue garage only wants to see you once. They don't want repeat business like an honest garage does, they want to scam you for as much money as they can and never see you again. They'll change all the things that they tell you need changing, but most of them didn't need changing in the first place.

Here's an example. A Range Rover 4.6 HSE had massive overheating, rough running, and coolant water loss. This car had been at a garage for three weeks, the invoice showed head gasket replacement, radiator repairs and a new fan motor. After I checked it, NONE of these things were changed, only cleaned up. The garage in question is a tiny backstreet 'bodge it and leg it' type. I called them and was told to fuck off and mind my own business.

The client apparently went there because the dealership she normally went to wanted to charge her an exorbitant amount, so she went to the first place she came across nearby. After it happened she contacted Trading Standards, who issued proceedings against them, so they shut down and re-opened one week later under a different name. She never got the money back from the garage but luckily she paid by credit card, and they stopped the payment. Then we had to do the job properly, and charged her some 50 per cent less than the dealership garage quote. There are some great independent garages, some even better than the dealerships, but be cautious: not all garages are the same. If in doubt, pay on credit card, as it might be your best chance of getting your money back.

Some garages are dodgy, others try it on, but some are just not very good at their job. I sometimes get cars come in and I wonder which of these three categories the car has just come from. Once I had a BMW 525I, year 2001, come in, with the engine management light on, transmission in limp mode and no power. This poor old gent owner went to a main dealership who quoted, wait for it, £4,950 plus VAT for a new gearbox! Sensibly, he flew out of there, went to a local independent garage who did a scan code, charged him £90 plus VAT and told him he needed a new gearbox and programmer, which would cost £3,850 plus VAT.

He was then recommended to me. I scan coded the car, noticed the alternator charge rate was very high, 16 volts, so replaced the alternator. Total cost? £198, plus £125 fitting, plus VAT. Result? No more transmission problems.

Open your eyes so-called mechanics, it ain't rocket science!

The point of what I'm saying is, how many of us would've taken the first quote? You reckon that if it's a main dealership, they know what they're talking about, right? So, quite a few of us think. And of those who had the balls to say no, I'm going to get a second opinion, how many would've taken the next offer, which was a grand cheaper than the first quote?

It takes some balls to get a third and fourth opinion, not to mention a lot of time and effort when you just want to get your motor running again, but sometimes it pays off big time. So when you're faced with a five grand bill, think how much money you could save compared to the time it takes to shop around. Saving a grand, well that's a fortnight's work for a lot of people and we're only talking a few hours to take it to another garage. This guy saved himself over four grand – think how much you can do with that or how long it would take you to earn that. This guy did it in just a few days, so he was well happy.

The one bit of good news is that most unscrupulous garages will get caught out eventually. A client once brought in a beautiful looking E-Type Jaguar, 1978, V12 and complained of poor performance, a fuel smell and a rattling when he went over bumps. The previous garage he went to was proving rather expensive: every time he took the car in for something the minimum bill was always £200 or £400. When I looked under the car I was literally shocked to see that a garden hosepipe was being used instead of fuel piping! The outer layer of the hose was literally melting away from the corrosive action of the petrol and was leaking badly.

The client checked his invoice from three months previously and saw that he had been charged £250 plus VAT for new fuel piping from the previous garage. On top of that the plugs were worn and old, the air filters were partially blocked, the oil was dark and smelling of fuel, the front brake pads were worn away and there was excessive movement in the steering ball joints. The company was issued with proceedings, and duly refunded him £1,800. Only then did they close down.

To reduce the risk of being scammed:

• Check your oil before you take your car in for a service. If it's dirty when it goes in and dirty when it comes out then it hasn't been changed. However, be aware that in diesel cars the oil will get dirty again within about thirty or forty miles because you're dealing with a heavy hydrocarbon, so make sure that as soon as you get it back from the garage you check it before it starts to discolour. Have a good look under the bonnet before you take it in for a service and look again afterwards – does it look exactly the same? Are the battery terminals cleaner than they were?

• Look at the wheel nuts. Do they look like they've been taken off recently? When you take wheel nuts off you should really put a bit of copper grease on the bolts. If they still look dirty and clearly haven't been off in ages, then obviously they haven't checked the brakes. Have a look at the door hinges. When you have a full service the door hinges should be greased or have the application of a clear lubricant. If this hasn't been done, or you find dirt and grease on the steering wheel, then they haven't cared for your car and this may set alarm bells ringing.

• Go in with ammunition. Say you would like a written or verbal quotation of what needs to be done on the car, that you want assurance that nothing will be done without your permission, then ask for the old parts back and ask for the choice of genuine parts or after-market parts. If you do that and show the garage you have a little bit of knowledge and understanding of your car, you're less likely to get screwed over. Don't go in like Bertie Big Bollocks and try to pull the wool over their eyes, because they'll see right through you and think 'Great, we've got a right mug here' and they'll try to teach you a lesson. So don't overreach, just be confident in asking the right questions, not giving them all the answers.

Garages are like any service that requires specialist knowledge. We're like dentists: you come to us when you're in pain and we'll tell you this needs coming out, that needs taking out and you don't know any better, you trust us because we've got a drill in our hand and say we know what we're talking about. So unless you become a dentist or mechanic yourself, you can never be 100 per cent sure you're not going to be scammed, but bear in mind all the things I've said in this chapter and you'll massively reduce the chances of getting shafted.

Remember, it's your money and they want it. You choose whose pocket it goes into and if you have any doubts at all, keep your wallet in your pocket or your purse in your bag.

You hold all the cards, so don't be afraid to play 'em.

CHAPTER ELEVEN
BANGLADESH

It's a long way to Tipperary, but Bangladesh is even further. It's funny, ain't it, but some time or other you get one phone call that changes your life. Mine was in 2006. I was working at the garage as per usual, nothing out of the ordinary, a few MOTs, a car that needed brakes, run-of-the-mill head gasket for me 'cos as per usual I was about to blow my top...

One client kept calling asking the same daft question: 'When do you think it will be ready?'

'When it's finished, Ball Brain!' I thought.

I mean, if I knew exactly I would always tell the customer, but cars are funny things at times and a simple job can turn into a right project. This cunt had bought a car in for a new driveshaft only for us to find that the whole assembly had been stripped. To add to this annoyance, the usual day-to-day running of the garage fell at my feet: ordering parts, chasing suppliers, paperwork, invoices etc. It's all building up around me and then the phone rings again.

'If it's that twat about his car again I'm gonna fucking kill him!' I think. I'm about to shout, 'No, it's not ready yet!' when I compose myself and put on my usual telephone voice, saying, 'Hello, Bernie speaking, how can I help you?'

'Hello, is that Mr Fineman?' came the words in my ear. I didn't recognise the voice but after confirming to him that he was speaking to the one and only I let him continue.

'Hello, Mr Fineman, my name is Dimitri from Raw Television, and I would like to come and see you about a new television programme that we are making, and we believe that you are the right person to be in it.'

I'm sat there taking this in and thinking that this geezer must think I'm a right grandfather clock if he thinks I'm falling for this wind-up. It's probably a mate's mate having a laugh, course it is. Before he could finish I just hung up. Fucking idiots I bet it's... Before I could think of the culprit the phone rang again.

'Hi Bernie, it's Dimitri.'

The way the day had been going I had no time for it, 'Look, phone some other cunt – offer them the job!'

He phoned again and this time I answered in my best Queen's English, taking the Mickey.

This guy was pretty convincing though and he knew that I thought it was a wind-up, 'Look Bernie,' he persuaded, 'why don't you call me back and you'll see this is for real?'

I took his details down, hung up, stared at the phone for a few seconds, then panned around to view the garage to make sure that no one was looking and laughing before dialling the number. I was still convinced it was my mate and was ready for the laughter in the background. I pushed the office door closed as the phone began to ring at the other end.

'Good afternoon, Raw Television,' an attractive voice

answered. This was definitely a moment to think of a good apology and eat humble pie. My mind began to race. You idiot, you idiot, I thought. 'How can I help you?' said the Raw Television lady. I gave my details to her and she said, 'Ah, Mr Fineman! Dimitri is expecting your call. I'll put you through.'

After apologising to Dimitri he gave me the details of a show he was producing for the Discovery Channel. Basically, he wanted a no-nonsense cockney mechanic, i.e. me, to star alongside a guy called Leepu Awalia, a talented Bangladeshi car designer who had no mechanical skills.

I don't know if you remember an advert for Peugeot set in India, where this young guy sees a Peugeot and wants it? He looks at his car and it's a heap of crap, so he sets about hammering all the panels on his vehicle and even gets an elephant to sit on the bonnet to crunch it into the shape of the Peugeot he wants. Anyway, they say that advert was based on Leepu, a car designer working out of a shack in Dhaka, Bangladesh.

The format of the programme would be that I would fly out with the crew to Bangladesh, Leepu would design a car made from an old banger, and I would have to get the thing to be mechanically viable. As you can imagine the conversation lasted a considerable time but it was agreed that I would meet them at a hotel in Borehamwood for a casting interview.

Now this sort of thing doesn't happen every day, and certainly I never expected it to happen to me. I had given some technical advice to Endemol Television on a previous occasion and apparently it was this encounter that had led to the call. You see, you just never know who you're talking to or who's watching; that joke or action you take could literally make or break you without you ever knowing.

The call had left me reeling with excitement and I called Lisa to let her know. She was as excited as me, but being realistic, she thought that since it was only an interview, would it come to anything?

Well, the time came around quickly and when I arrived there was a television crew and Dimitri, who turned out to be a lovely guy. 'Bernie,' he told me. 'Just ignore the camera and be yourself while I ask you some questions.'

My eyes rolled, the camera rolled, and at the end of it I was offered a contract to star in the show. Double the money I was on, it sounded fantastic with an opportunity to make another series afterwards. There were several conditions and the hardest was that we had to deliver a car a month or Discovery wouldn't pay and the show would close.

The small print didn't seem to matter to me and I had no agent or adviser – I just had to get Lisa's approval because I would be out of the country for a minimum of ten weeks. Later on it was the small print that lost me a lot of money, but for the moment all was cushty. By the end of the year we'll be *millionaires*, Rodney, *millionaires!*

When I arrived home Lisa and I chatted into the small hours. We had not been apart since first meeting and although she was saddened that I would be away for so long she said, 'You have my blessing to go, Bernie.'

I hugged her close, telling her, 'I'll buy you that Merc I always promised you.'

She just hugged me before saying with a smile: 'Just be good Bernie, or I'll have your guts for garters!'

That was that then. In a few short days my life had turned from the mundane to the unknown. Was I a little scared?

You bet I was.

The following morning I called Dimitri. *Bangla Bangers*, as the show was going to be called, was on. I resigned from my job with immediate effect. I don't like letting people down, but the contract stipulated that I had to fly within seven days. So I had seven days to get my jabs: yellow fever, typhoid, diphtheria, tetanus top-up, you name it. I also needed some lightweight shirts, headache pills, toothpaste, new undies, mosquito repellent, Diacalm and balm-coated bog-roll. No Jay's toilet paper sandpaper for me!

When the day came to go I was sadder than I had ever been, but excited at the same time. Lisa took me to Heathrow airport where I met up with all the crew. Hiding the tears I walked through the gate looking back at Lisa, still in love like the day I had met her, before walking with the crew into the departure lounge. For a few moments I was lost in the crowd and thought of Dad, saying to him in my mind: 'I'll make you proud Dad, I'll make you proud.' The noise and bustle of the departure lounge brought me back to earth, and there were the usual searches and questions, such as, 'Did you pack your bags yourself, sir?'

I've always wanted to turn around and point at the nearest person and say, 'No, it was her who packed it!' There were more daft questions: 'Are you carrying any scissors, sir?'

And I replied, pointing to my bald pate: 'Yeah, with hair like this I need 'em.'

I just had to listen to the usual bullshit.

It was only when my arse got onto the plane that the reality hit. Fuck it, I'd left my job, I'd left Lisa behind, and the words 'If there's no car produced in four weeks there's no pay' kept bouncing across my mind. They meant it as well. But was it possible to do it? I was used to hard work but everyone has their limits. I knew a little about my co-host Leepu and had seen some pictures but I had not met him. Were we going to get on?, that was the question. Unbeknown to me, that question was part of the point of the series.

The plane engines started to wind up and as we taxied along I knew one thing: there was no going back.

After a bum-numbing thirteen-hour flight we arrived at Zia International Airport in Dhaka, and went through their customs to face more of the same daft questions. Some bloke got arrested for bringing in more than two hundred sheets of bog-roll – what an arse-wipe! Anyhow, jokes aside, passports were stamped and I'm off into the arrivals area where I knew Leepu would be waiting.

The crew had so much gear to gather that it took a little time before we walked through into the great unknown. The heat hit me like a roasted cricket bat as I walked through the exit doors, and immediately I saw Leepu standing in front of me and was aware that the camera was already rolling. It's funny but you never appreciate how much footage is shot for a single programme.

We hugged and introduced each other before being led to a car, where we were allowed to remain chatting for ten minutes as we travelled along the streets, before being separated again.

'I'll see you in the morning Bernie,' Leepu said in an Americanised Bangla twang.

Little did I know that this fat, one-sandwich-short-of-a-picnic crazy, whisky-drinking little Bangladeshi man would change my life. He went his way, and I left for the hotel in the company of the production crew.

What a hotel! It felt palatial and the staff treated everyone like royalty, bags carried, absolutely nothing was too much trouble, but this was a far cry from what I had seen outside. This regal establishment could not hide the real poverty just a course of bricks away where people live a fragile existence.

The evening was spent as you would expect: nice bath, dinner, a few drinks in the bar and then bed, ready for an early start. I still did not know what to expect but that was half of the challenge and I was ready, or so I thought.

So it was up bright and early the following morning, for a quick brief from the production crew. Camera ready, soundman ready and this is it, we're making a TV show. I travelled by rickshaw through the streets, which was life-threatening in itself. The rickshaw just behind me was shunted by a lorry, but luckily no one was hurt as far as I could tell.

I arrived at Leepu's garage and he invited me in to have a look round. Fuck me, there were less tools in there than in a brand new transit: just a few hammers on dusty shelves, an acetylene torch with no safety device – one mistake with that and it really is Bangla Bangers. There were dodgy electrics, and the little power tools that were to hand had no plugs and were wired directly into the sockets! The smell weren't too clever, as an open sewer ran through the premises. If that alone wasn't bad enough, the heat beat down like nothing I'd encountered, humidity was just off the scale.

Still, I thought, this is what we had and we just had to get on with it. At least I didn't live there, unlike Leepu and his family. I just kept thinking about Lisa and my home in Borehamwood and how lucky we were.

The first thing to do was to acquire the car, and cars in Bangladesh are really expensive, even fucked-up ones. Leepu has his designs in his head so I did not have a clue what the car would look like. If he gave you a plan on a piece of paper it would be blank, literally.

The bloke was off his head!

After trawling around several car dealers (no courtesy coffees offered here), Leepu found the car he wanted: a two-door Toyota Sprinter. This thing looked fucked, the kind of motor that if you took it to a scrappy you would have to pay them to take it away. He asked me what I thought of it and I told him straight, 'It's a piece of shit!'

The greatest shock, though, was the value of the car. Sixteen hundred quid was exchanged before it was loaded and taken back to Leepu's garage. Sixteen hundred quid and it didn't even run! Still it had a full service history... behave! On arrival back at the garage it was pushed in by Leepu's brother and co-workers, and we started to strip it down. The clock was ticking. The Discovery Channel wanted the first series, or they'd have our blood.

With all the major body components off, the car was tipped on its side. No health and safety here, no fancy jacks or body roller. No, chuck it on its side and get your head in. When we had a look at the underneath it was truly shocking, rotten as a pear, and I mean *rotten*. In the UK the car would be gone, but these lads just get stuck in, and over the course of two days they chopped the rot out from underneath and, using handmade panels cut from sheet steel, they remade it. I was amazed by what they managed to achieve.

We left them to it and went looking for new wheel rims. By the state of the car it seemed a bit early to be thinking of things like that, but we needed the parts. Leepu led me all over Dhaka until he finally found some alloys he liked. The price was good at around £300 but God we needed so many parts. We eventually had to wait for two hours for the wheels but at least we had them.

Now I'm Jewish and keen on saving money, but Leepu certainly taught me a thing or two about how to barter. The shopping spree lasted well into the night with row after row of small auto parts vendors who had everything you could imagine offered for sale. At one point I was buying shock absorbers by candlelight due to a power cut. I called it romantic and in a way it was.

The final purchase of the day was for the sheet metal. The plan was to buy two gauges of sheet steel – a thinner gauge for bodywork and a thicker one for structural work, and every panel was to be made by hand. The steel was loaded onto a rickshaw as well as into the arms of both Leepu and me. The chap riding the rickshaw had some power to shift our fat arses and all our cargo, bless him. On arrival we worked until late into the night, which would become the norm, because double-money meant double-hours. Still, we were both fired up, although we realised it was going to be harder than I thought, much harder.

The following morning Leepu started dancing around the car with a bit of metal in his hand flapping it up and down. Clearly the design was being imprinted in his mind. In his own words: 'You have to get into the design to get it out.' He gave me a rough idea of what it would look like: a low slung, wide two-seater. It sounded good but still I had no visual idea, and the mechanics were down to me. The car's body was stripped of paint by using nothing other than old hacksaw blades as scrapers... bloody hard work, no power tools here. I couldn't believe it, there was no electric or compressed-air powered grinder, no chemical paint stripper, just manpower.

The suspension needed to be lowered. This is done by reducing the length of the springs and shocks, but there are limits. Without any consultation Leepu hacked seven inches off the springs and the shocks. That's too severe and, frankly, I was livid. What the fuck was I doing there trying to help him if we didn't consult each other before charging ahead?

Seconds were ticking by and I really was starting to feel the strain. Time was money and so, against my better judgement, the suspension was refitted to the car and the new wheels fitted on massive six-inch spacers. Leepu wasted no time in chopping the bodywork, lowering the roof, cutting through all supporting pillars with a hacksaw and taking a section out of each of them, before welding the roof back on. I must admit the car had a nice angle to it, but at four in the morning anything would have looked good.

We had three weeks to go before the car was due to be unveiled as the star attraction at the Bangladeshi car show. It was apparent to me that the car needed a bigger engine, so Leepu's brother Dheepu took me to a car parts bazaar, leaving Leepu to carry on with the work on the body. At the bazaar everything you could wish for was there but it was an engine that was required, a five-speed manual, 1800 or two litre, and it had to be fuelled by carburettor for simplicity.

Dheepu led me down these alleys, reassuring me that he knew a place where such a unit could be found. I was expecting another small outlet with a few engines, so nothing could have prepared me for what we found. On entry to the warehouse we saw that the place was full of engines, hundreds of them, all salvaged from cars from Korea and Japan. After a bit of a search we found the right unit – an 1800.

But did it work?

Turning it over by hand was one thing. I had forgotten, though, that this was Bangladesh and if you wanted to see something working, no problem. Two men were summoned and they picked it up, by stringing it up using a cam-belt over a pole. This reminded me of the time I dropped that Vauxhall engine that jammed in the engine bay, and my dad had to come and pull it out for me. Still, there was little time for memories.

The chosen engine was dumped on the pavement, where a large crowd had gathered around us. One of the fellas turns up with a big fuck-off battery and a cup full of petrol. I couldn't believe my eyes! This geezer stood on the battery to ensure a good connection as the terminals were loose, and poured the petrol in, as his colleague shorted the ignition and started the motor.

A small child was looking on. His face was just inches from the front pulleys, and it scared me that if anything had gone wrong his face would've been torn off. The engine chugged into life and there was a massive roar from the crowd. I could tell it was a good one – those years in the trade give you good ears for a lemon and this was a fucking peach.

The obligatory rickshaw was summoned and both myself and Dheepu made our way back, engine and all. On arrival the amount of bodywork that had been completed in my absence was astounding. New structural braces had been fitted across the original cross-members and new hand-beaten wings had been welded front and rear. It was a fucking ace job. Leepu was ecstatic with the engine, and I had taken an engine mounting with me, so I knew that the unit would fit in the space with very little modification needed.

Leepu continued to work solidly for three days on the bodywork but I could not get access to the mechanics and I was getting fucked off. This wasn't just a car, it was my mortgage, my wife's home! At this point I might just as well have just stayed at home. Days went past with me getting nothing done and words were being exchanged. I got the engine in but, disregarding my needs to form a bonnet scoop, Leepu welded the thing shut. I watched in horror as they managed to ignite the engine underneath the closed bonnet with the dodgy acetylene torch. It was panic stations to put it out.

I had to source a new radiator and spent hours waiting for the part, only to arrive back and find that the radiator needed to be modified – one of the outlet ports needed moving. My temper was starting to fray. Leepu couldn't care less as long as the bodywork was done; the trouble was that he was taking me for a prat – a bad mistake.

Yet again I travelled with Dheepu to a local specialist and in time-honoured tradition the radiator was modified on the pavement, soldered, pressure-tested, the lot. Guess what? On my return he still wouldn't let me gain access to the engine and we ended up having a stand-up row. Fuck 'em, I thought.

I went back to the hotel, returning in the early hours of the following morning and found to my delight that Leepu was asleep. Great, I thought, the cunt's in bed so let's get to work! Radiator fitted, coolant added – and the engine started. It was running a little rough but a quick hand on top of the open carb and any crap in the fuel chamber was sucked, banged and blown out – it took me back to my lads' holidays to Thailand!

Then I worked bloody hard on the engine to get it running right but I wasn't in for any praise, far from it. Leepu, having woken up, just stood there with his hand on his chin. I tell you I was close to putting my fist on it. I was unhappy with the mechanics of the car, especially the suspension, which I considered to be a potential deathtrap.

The following day Leepu was off buying the finishing touches, steering wheel etc. It was while he was away that I spotted another potential killer. The prop-shaft (a huge revolving shaft that runs under the car from the back of the gearbox to the differential between the back wheels, thus transferring the power) was not aligned with the gearbox, and the constant velocity joint (positioned part-way along the prop shaft) could have exploded, sending the whole prop through the car! To rectify this we managed to raise the gearbox within the body to align the shaft. At the end of all this work I was fucked.

Leepu returned with his steering wheel and a piano accordion which he started playing. Now if you can't understand what's being sung it doesn't matter if it's in tune, but to me this instrument always sounded like they've dropped an anvil on someone's toes. In any case there were more important things to talk about and it was my turn to hurt his ears.

I explained about the difficulties we'd had with the prop-shaft and my feelings regarding the suspension, e.g. the lack of it, but Leepu appeared oblivious. The car had many mechanical problems but the test drive was looming. For a start the petrol tank couldn't be filled because the filler neck was obscured.

Shouting and swearing appeared to be the order of the day with both of us getting in each other's way – that was the bad news. However, in the early hours of the following morning, while the roads were quiet, guess what? The beats hit the road.

There are no worries here about insurance, MOT, road tax, you just get out and drive. Leepu flew down the backstreets of Dhaka with me sat in the passenger seat, and the few locals who were about at that time of day just looked on quizzically. The car looked and went great but my God it was rough to sit in – the rear end was hopping about like Bugs Bunny on acid.

The result of the test was yet another row between me and Leepu. Basically I dressed him down for not listening to me with regards to the suspension and he went off in a huff after I suggested lifting the rear slightly. Leepu sarcastically claimed that I wanted a monster truck. Saying that lit my fucking fuse and I told him that if he had spoken to me like this in England I would have fucking killed him.

I was not happy.

We made up the following day but the pressure on us both was building. The crew wanted the footage to cut for Discovery and time was running out. We managed to compromise but it was still tense. But at least I was allowed to do some fucking work.

This time around Leepu worked on the body and I worked on the mechanics, and at last we were working in some form of sync, but, fuck me, we had only days left. Leepu's lads had covered the whole car in body filler, it looked like cement and it needed sanding down, applying the primer and paint, let alone doing the detailing. With hard work, and I mean *bloody* hard work, the sanding and paintwork were completed.

I would never have thought it could be done. *Never.* Just goes to show what can be achieved when a group of guys put their minds and determination to it.

The detailing was done including putting in modified headlights that Leepu cut with a saw. The end product was truly amazing, in fact it was more than that – it was a miracle. With just sweat, a few tools, arguments and persistence, the impossible had been achieved. The only question now was: How were the public going to perceive it?

On the day of the 'Big Reveal', like a bride to her wedding we were late, but the polite and dignified people of Bangladesh stood in readiness as we arrived at the show in the nick of time. Their reception for us and their anticipation was incredible, totally different to any crowd you would get in the UK – they were so excited.

I gave a speech to the crowd, most of whom cheered without understanding my native tongue. Or was it just my Cockney accent? To them it didn't matter, for they knew what we had achieved. The crowd started to cheer as I told them: 'It gives me great pleasure to reveal Leepu's creation.'

It was awesome. It looked like a Ferrari and as the wheels rolled, so did the tears down Leepu's face and mine. It wasn't just a car; it was a collaboration of culture, religion and faith.

I looked round at the crowd and thought of my dad smiling. *We'd done it!*

Well, we had done it this time. Tomorrow was another month and the Discovery Channel wanted another car.

Well fuck my aunt.

A month later and another car was done, Discovery had their series and I was off home, thank fuck. I couldn't wait to get home to my comfy bed, my beautiful wife and family, but as we boarded the plane something incredible happened. I realised I didn't want to leave. It had been such a powerful, all-consuming experience, the people and the place had just been incredible and I didn't know if I would ever go back or see Leepu again.

We missed our inter-connecting flight so had to spend the night in Abu Dhabi and I didn't sleep a wink. All I could do was think about all the friends I had made in Bangladesh and wonder if they would be OK. It was so intense, I just couldn't get it out of my system. I know that when I left Bangladesh a part of me stayed there.

Words do not do justice to the conditions in Bangladesh. It was forty degrees in the shade and we were working from a tin shack with an open sewer running through it. To make the stench worse, next door they were decapitating live chickens! And all the time me and the crew were suffering from the dreaded Delhi Belly, but the only toilet was just a hole in the ground running to the sewer.

They had no toilet paper so you'd do your business, then dip your hands in this bowl of water and wash your arse with your hands, then wash them in the same bowl of water. You can imagine the germs and infections swimming around in it! We all lost at least a stone out there and one of the guys in particular was really ill – when he got home he looked like he'd just come back from Belsen, not Bangladesh.

All we ate for two-and-a-half months was curry. Now, I love a good curry as much as the next man, but three meals a day for ten weeks takes its toll, and the first thing I did when I got home (after kissing my wife) was eat bacon and eggs. The problem was that whenever we ate, we were ill.

There was a doctor in the hotel, though, and he was making a bloody fortune selling us his diarrhoea tablets. Eventually he took pity on us and told us not to keep taking the tablets as eventually they would ruin our stomachs. By the way, if you ever get the runs, all you want to do is get a can of full-fat Coke, open it and let it sit there until it goes flat, then drink it. Coke stops the runs straightaway: seems it kills all the bugs in the stomach. So that's what we'd do: buy loads of cans of Coke, open them up and put them on the side then drink them throughout the day. Leepu told us the best thing to do was to eat in the most popular restaurants in town. That's because they had the highest turnover of food, so you knew it was always fresh. We hardly had a problem after that.

Working in such conditions and being in a foreign country you get to know the crew really well and we all became good friends. Many nights we were working and filming until three or four in the morning, but when we did get time off we'd go out for dinner, take walks or go and see the sights. One night we finished and the rest of the crew went to bed except me and the sound recordist Martin, so we decided to try out a new restaurant that had just opened up, that was named after the Bollywood actor Sanjay Kapoor.

It was about a twenty-minute drive away, but well worth the taxi ride as the food was fantastic and we had a great night. On leaving we hailed a taxi and asked to go back to the hotel. After a few minutes he says 'petrol' and pulls off into this garage, well off the beaten track. You couldn't even call it a garage, it was just someone's house with a petrol pump in the front garden. The guy gets out and starts to fill the car while a group of blokes are stood around in the front garden.

Martin turned to me and told me he didn't like it, he was sure something was up. These guys keep coming over and looking at us through the window and when the driver gets back in the taxi it won't start. When the light outside the shed turns out I say, 'Right, we need to take the bull by the horns.' I get out of the car, pull my phone out of my pocket and show it to everyone.

'Get me another taxi now or I call the police,' I told them. Knowing they'd been rumbled at whatever scheme it was they had going, they found a car round the back and took us straight home, but all the while Martin and I were absolutely shitting it.

Thankfully that was the only bit of funny business we encountered, which was lucky because you do hear a lot of horror stories about Westerners travelling to poor countries in Asia and they are sometimes seen as easy money. But everyone else was so friendly. We were there during Mela, their New Year, so we joined them as they danced and drank in the streets. We had our faces painted, and the atmosphere was absolutely electric.

Being such a poor country, if something broke you had no option but to repair it, and so the result of this is that you have a highly skilled population. You would see young kids in the street, sat on the floor, legs apart, holding a car door that had been smashed up. They'd sit there for five or six hours with a hammer and dolly getting all the dents out of it to get paid probably a few pence.

As for the guys in the garage, if there was a radiator leak they would put solder around it, but to pressure-test it they'd fill it up with water, put their mouth round the top of the outlet and blow into it non-stop for a minute, or a minute-and-a-half, without breathing, putting pressure through the water to make sure there were no leaks. It was an amazing thing to see.

For me, it was the closest I will ever get to time travel: it was like going back to the 1950s when I first started in the garage trade. There were no cranes, so when we had to lift in this engine that weighed about 400 lbs, we did it the old-fashioned way. I got a bar and we tied it to the engine with all these old fan belts. There were three guys one side, me and Leepu on the other, and that's how we got it in and out. No steel toe-capped boots there – if it dropped on one of those guys' feet that was it, it would probably have to be amputated. With the barest of hand tools and no ramp, as I say, it was like going back in time. I never thought I'd work like that again, it was just an incredible experience.

Going back in time also had its downsides, though. One day we went out for lunch and I was sat down eating a curry when all of a sudden something cracked in my mouth. At first I thought it must've been a bit of chicken bone or something but I soon realised it was one of my teeth. My dentist in England later told me it was my Upper Left 4. I spat this bit of tooth out and thought not much more of it, but as the afternoon wore on I developed the most raging toothache. I've never felt anything like it, just throbbing constantly in my mouth.

When I was a kid if we got toothache we'd use oil of cloves, but there wasn't much chance of me finding any of that in Bangladesh. I took whatever tablets I could get my hands on but it was no good, the tooth was cracked in half and the nerve was exposed. The pain was indescribable and I was completely debilitated, so the production crew went on a desperate search for a dentist, eventually finding one about a fifteen-minute drive away.

When we arrive, it's like a small shop. First there's a waiting room, then when it's your turn there's a door you pass through to go out the back where the action takes place. I was quite calm sitting in the waiting room with the director and our driver, but nothing could prepare me for the horror that awaited me through that door. It was an open-plan room, nothing sectioned off, just six chairs with people sat in them having their teeth done.

I stood there looking at these six poor bastards writhing in agony as the dentists worked on them. The equipment was like something out of Victorian times. There was no electricity, so instead they used a treadle to operate the drill, with all these pulleys hanging off the ceiling. On the table next to them was a tray of instruments and they just looked filthy.

The third dentist along beckons me over and, to my horror, I feel my legs walking towards him. I sat on the chair and he took a look inside my mouth. Now, I'm not afraid of no one or nothing, but as he looked in my mouth I looked around at the guy sat not two feet away from me, moaning with agony as he's having his tooth drilled. I thought, 'No way, I ain't having this,' and as he reached for one of his instruments I shot up out of the chair and legged it. Pain or no pain, that seemed like a fate worse than death. Genuinely, I thought the tools we had back at the garage, negligible as even they were, were a damn sight better than what those dentists had, and they were probably cleaner too.

As I emerged from the room the director looked at me – my face must've been ashen. I just shook my head and said, 'No way.' The painkillers had finally numbed it a bit so I went back to the workshop and got on with things. An hour later, though, and the pain starts coming back with a vengeance so I decide to take matters into my own hands.

On one of the shelves was an old bottle of whisky, so I got it down, emptied a triple into a mug and necked it. A few moments later, as I felt the whisky and Dutch courage kick in, I grabbed a pair of pliers and put them round the tooth. It was a little loose already, so after a few wiggles I just went for it. There was a momentary rush of pain and then it was gone. The guys just looked at me as if I was fucking nuts! Some of them said they didn't know how I had the guts to do that, but I said if they had the same extraordinary pain as me and if they'd seen that butcher's shop of a dentist, they'd have done the same, believe me!

When I got home I was overjoyed to see my family, but there was no time for recuperating, I just went straight back to work. I didn't really think any more of it until one day we got some DVDs in the post. Raw Television had sent me the DVDs of the series to watch before it went out on the box.

That weekend we invited the whole family around to watch it. I've never been so nervous, but I just knew that it was going to be a good show. We all gathered around the TV and there wasn't a dry eye in the house. To see yourself on television, for a bloke like me, a poor boy from the East End, was just amazing, the stuff of dreams, and for your family to be there and for them to tell you how proud they were of you, I'm not ashamed to say that I cried too.

A couple of months later the show went out on TV as the first series of *Chop Shop* under the name *Bangla Bangers*, and a few nights after that my wife Lisa and I were in a Japanese restaurant on the Finchley Road. This guy came over and said, 'I saw your programme on TV, absolutely fantastic, can I have your autograph?'

Well! Did my shoulders swell and did I have a smile?!

I saw a glint in Lisa's eye too, she was so proud of me, and we exchanged smiles. Fuck me, I'm famous! Since then barely a day has gone by that I haven't been recognised by someone. For the most part it's been a wonderful experience, everyone is so friendly and pleased to see you, you feel like you've made their day when they recognise you.

It's amazing the places I've been spotted – you just don't realise how many people have seen these programmes. Lisa and I went on holiday once to Turkey, and just walking up to passport control I was getting calls from other holidaymakers, saying things like, 'Hey Bernie how are you?' Though, to be fair, I did happen to be wearing a *Chop Shop* sweatshirt!

There were the usual huge queues at Heathrow when a couple of the security guys spotted me. 'Hey,' one of them said, 'It's Bernie from *Chop Shop*!' Everyone turns to look at me and I'm mortified. They beckoned me over to one of the scanners that wasn't in operation: a quick frisk and we were straight through, we bypassed the entire queue. Now I was even more embarrassed – hundreds of pairs of eyes looked at us as we jumped the queue.

At that time I hadn't realised that the show was being broadcast in ninety-two countries, so when I got to Turkey I was still getting recognised, there was no escape! I've been spotted in Finland, South Africa, all sorts of places. When I've been shopping in Asda people have come up to see what I've got in my basket, and they're probably surprised to find bacon!

A few weeks after the first episode went out I got a phone call from Raw Television saying how delighted they were with the series, that it was one of Discovery's most popular car shows ever, and that they wanted to do another one. This would mean another year of my life filming an eight-episode series. I just burst into tears. I still didn't think I had found my niche in life, so to hear that I was wanted was an incredible feeling.

This time, though, things would be different. They wanted to take Leepu out of his comfort zone and take him somewhere where he could stretch himself and create even more amazing machines. So, where better than my manor? The East End of London, where there's a large Bangladeshi community, but with modern technology as well.

Before we knew it Leepu and his family were on a plane to England, they sorted him a house and we were making *Chop Shop: London Garage*. For goodness' sake, we barely came out alive after two months together, how would we survive another twelve, I wondered?

I remember Dan Korn, then head of programming at Discovery Europe, saying to me once, 'We need more language.' By that he meant swearing. Well, I didn't need asking twice with Leepu around! I did have slightly mixed feelings, though. I was still doing my consultancy work and building up a new garage, and the new series would put a stop to that. A year out of a business can be a long time, especially so soon after coming back from a ten-week break in Bangladesh. But in the end I decided these sorts of opportunities don't come around very often, and my reputation was good in London so there would always be work out there somewhere for me. So yes, I decided, let's do it.

When I went to Bangladesh I felt I needed to prove myself to the locals, who might be thinking, 'Who is this white guy coming over here to tell us how to fix cars?'

And it was the same for Leepu when he came to England. I think he felt he had to prove himself even more and that he would be judged by British standards rather than Bangladeshi standards, which put a lot of pressure on himself to raise his game.

Sometimes it would get too much for him, and his mind would go blank. He'd get an idea in his head, start on it, then walk away and leave it. We didn't know what he was thinking from one minute to the next. For the first three weeks he'd roll in at 9.30, 10 o'clock and go home at 3 pm. He'd say, 'My mind's just not with it, I can't focus.' But we had a show to make and the mechanics had to carry on. It took him a while to convince himself that what he was doing was right, but then he came up with some fantastic creations.

Leepu soon grew to love London. There was a big Bangladeshi community so he didn't feel like a total outsider, he loved the fact he could get a drink twenty-four hours a day if he wanted to, and he loved our burgers. Whereas we lost weight in Bangladesh, Leepu went home having put on twenty pounds in that first year in London!

The car I am most proud of from that first run of *London Garage* is the one Leepu said would never work, the one we called 'The Angry Frog'. It started off as a clapped out, beaten up piece of fucking shit Mitsubishi Pajero four-wheel drive. It was an old smoker that drank diesel like you or I would drink water, but I wanted to do something different. I wanted it to run on old cooking oil.

There were companies out there that sold conversion kits for £2,000 but we only had a budget of £4,000 for the entire car, so there was no way the production company would sanction it and, anyway, it wouldn't make great telly just buying the kit and bolting it on. So I decided to design and build my own. I went over the design with Leepu and, ever supportive, he said, 'It'll never work, we'll be pushing this car.'

All the way through he doubted it and doubted it and doubted it. In the end I just told him to fuck off and concentrate on his own design, which looked shit, and to leave me to it.

For the grand total of £10 I built my own conversion kit for this car, and what did we get it to run on? Well, being only a short trip from Brick Lane, it had to be curry oil! So we went round all the shops and restaurants with 25-litre drums asking for their old curry oil. I built my own strainer to get all the crap out of the oil, my own heater to heat up the curry oil before it went into the engine, and bought two cheap filters which I attached to an electronic switch to heat the oil up to 85 degrees.

Five days later, much to Leepu's amazement, the first time we tried it the engine started up on the button. The only downside was that when you stood behind it, you got really hungry 'cos all you could smell was fucking curry! And it flew like a rocket, without hurting the environment.

What could be better? Make yourself a great dinner, enjoy it, then stick it in your car and drive!

Just like *Bangla Bangers*, *London Garage* was hard work. There wasn't the heat and of course we had better tools, but there was also the pressure of upping our game and, of course, being a much longer series it was nonstop for a whole year. We had just £4,000 and four weeks to build each car and there was no room for failure: Discovery needed their TV show – it didn't matter how we did it, it had to be done.

We had a great big whiteboard on the wall and on it was written what we were going to do each day: brakes, steering, electrics etc. and whatever it said we had to do, no matter how long it took. There was no script, it was a case of 'What you see is what you get'.

The most stressful car to make was the congestion-buster. We got an old square sandwich van and we were going to convert it so that you could take it into the West End without having to pay the congestion charge.

Once we'd drained off all the fluids the first thing Leepu proceeded to do was start chopping it up, so a vehicle that started out eight foot long ended up just six foot long, then went down to five foot. This made everything fifty times harder for me and the mechanics, because by reducing its length he'd taken the whole backbone out of the car.

Every time we made something to facilitate seating in the car it wouldn't fit because the compartment was too small. It was so small that if you were anything over five-foot six inches, then your arse was sat on the brake lights and your feet were hanging over the front headlights, and in between we had to fit an engine, steering and suspension. The more he chopped it the more work it made for the rest of us to make the car safe, still run AND convert it to run on LPG.

In the end it turned out brilliantly, but the amount of arguing over that one was unbelievable. You can see it in the film, but that was just a taste, there was so much arguing there wasn't enough time in the programme to fit it all in! Dan Korn at Discovery definitely got his swearing quota on that episode. And that was without the biggest fall-out me and Leepu ever had.

Tempers were fraying and Leepu was constantly going on at me, saying things like, 'Your mechanical work is shit. Your mechanical work is holding me up. I don't want your fucking engine, we'll just push it instead.'

Eventually I snapped and I lost my temper like I've never done before or since. I literally lost it, the red mist came down and I grabbed him by the throat, pinned him up against the wall and proceeded to choke the life out of him. I wanted to hurt him bad. In that moment I could've killed him, and he could see in my eyes that I meant it. The camera crew stopped filming and rushed over to pull me off him. It was so serious that they told me to go home immediately and report to the MD of the company the next morning, and they were genuinely worried that the series was over. After all, there was no way Leepu was going to work with me again after that and rightly so.

It was two o'clock in the afternoon and I was too ashamed to go home: Lisa would be there and I couldn't face her, so halfway through my journey I pulled up by the side of the road and just sat there. My world had collapsed. All I could think was: 'You fucking idiot, you've ruined it! They're going to cancel the series. What will your friends and family think when you tell them what you've done?'

Eventually I drove home, but I couldn't tell Lisa what had happened, I was too ashamed. So I told her we'd been let off home early because we'd been working so hard recently and we'd got what had to be done that day anyway

The next morning I left at my normal time, 6.30 am, so that Lisa wouldn't suspect anything. I got in the car and drove to Sainsbury's car park in Watford, killing time waiting for the call. Eventually they ring at me at 11.30 am. I've got a meeting with the production team. This is it, I think, they're going to fire me. When I arrive Leepu's there already with them all waiting for me. They look at me and ask what I have to say for myself.

'What can I say?' I admitted. 'I'm deeply sorry. That wasn't me yesterday, I've never lost my temper like that before and I never will again.' I turn to Leepu, telling him, 'I'm truly sorry, from the bottom of my heart I am truly sorry. Please forgive me.'

Thank God, Leepu said he accepted my apology, we shook hands and it was never mentioned again, but it was never the same after that. We got on and everything but we never quite had the same bond, and to this day I regret that.

We finished the series with no more dramas on that scale, and when it went out, by all accounts, it was even more popular than *Bangla Bangers*. This time it wasn't such a surprise when Raw called to say they wanted to do another run of *Chop Shop: London Garage*, and again they wanted to up the ante. This time we would be making 'cars for stars' but we wouldn't know who the stars were until the day before we were due to meet them.

I thought this was a great idea and couldn't wait to hear who the first celebrity was going to be. A couple of weeks later I get a call saying I'm to meet Martin Kemp in a bar in Bethnal Green the following afternoon. So I turn up and we chat and I ask him what sort of car he wants. He says he's an East-End boy as I am, and he wants a proper 'gangster car'. Well, even if he had played Reggie Kray in a film, I could tell he was no gangster, he was in Spandau Ballet for fuck's sake!! And he had a handshake like a wet wank, but he said, 'If your cars are as good as your chat, we'll have a deal.'

So we go back to the garage and discuss with the production team what we're going to do, how we're going to create this gangster thing on the budget we've got. We decide to base it on an old Saab, but all I was worried about was: we've got this nice 2-litre front-wheel-drive car, what the fuck is Leepu going to do with it?

His mind is racing and he wants to do a mega design, which is fine, we can work with that, but he wants an ultra-powerful engine to go with it. Now what he wants and what's in the budget are two different things. Again, it was just £4,000 for each build. There are very few engines that fit a Saab, we could've fitted the 2-litre Turbo version, but it was so rare and expensive that wasn't possible.

All the time the crew said just make it look as good as possible. I took the engine out, changed the seals on it, cleaned it all up, made it look beautiful, tuned it, and blinged all the hoses by using those bright blue ones. Basically did what I could to make it look good, while spending as little money as I possibly could on it.

Then Leepu got his hands on it and, between you and me, it was a total fucking abortion. I was ashamed of that car. The way he did the back, fine, it was practical for putting dead bodies in, but it wasn't practical on the road. Mechanically I was very proud of what we did, I got as much power out of that engine as humanly possible on no budget, but nothing fitted on the body. Every time Leepu changed something we had to reinforce the body, so we ended up with doors not fitting properly, all sorts of problems like that. It was another one of those where Leepu's ideas went beyond the practicalities of reality.

Then there was the problem of Martin Kemp. He rubbed us all up the wrong way when he first came down to the garage to meet the team. After shaking us all by the hand he insisted on going to wash his hands, which came across as incredibly disrespectful. Then when he spoke to us he had an air about him that I didn't like. In my opinion it seemed like he felt as if he was talking down to us. We just did not get on.

After a couple of weeks the production team say they want

some 'jeopardy' in the programme, so they arrange a track day to road test the car. What we had to do was cut the top spring turret to make the car unsafe. When we turned up on the day we made sure the spring slipped as we took the car off the back of the trailer, at which point I said it wasn't safe to drive.

To be fair to Martin, I don't think he knew this was going to happen, as he'd brought his son along with him for the test drive. But of course he absolutely went off on one, saying we were wasting his time and that his son could've been killed. Once he'd finished and stormed off we took the car back to the garage and proceeded to make another modification to make it look good for the camera and made it safe.

But the production team still weren't happy – they didn't think Martin came across strong enough on camera, that he wasn't gangster enough – so they asked if there was anything I could do. So I made a few calls and got in touch with Frankie Fraser. I explained to him that we hadn't spoken for twenty-five years but that I used to do his cars and told him about the TV show. I said they wanted him to meet Martin Kemp. Of course, Frankie knew who he was – Martin and his brother Gary had played the Kray twins in the feature film – and referred to him as 'that fucking plastic gangster!'

I asked Frankie if he would mind being on the programme if we met him in the Blind Beggar for an interview and, if so, how much he would charge. He thought for a second and said, 'Bernie, for you, buy me a couple of pints and a bit of lunch, there's no charge.'

The crew were over the moon, I'd got Gangster Royalty to be on the programme and in lieu of payment they gave a cheque to one of his charities. Not that it really makes up for his crimes, but Frankie was a big supporter of several charities.

We meet up a few days later at the Blind Beggar. Although I had a lot more hair the last time I saw him, Frankie recognised me immediately and chatted as if it was only yesterday, then we take a walk around the East End. Even after all these years, and Frankie must've been about eighty years old at the time, people were still coming up to him in the street wanting to shake his hand and show their respect.

He came into the garage and shook hands with all the mechanics. He was so polite and warm with everyone. What a lovely, lovely guy. And he loved the car; he took one look at Leepu's boot and said, 'I could get four bodies in there, good work son.'

Later he was chatting about some of the things he'd done in his past, and this was on camera. He told a story about a job to rob some tomfoolery (jewellery) when the getaway car broke down. He suddenly shot me a look. 'You didn't fix that one, did you Bernie?' he asked. Fuck me, he might've been old and frail, but that look he gave me I'll never forget, I totally shat myself! It was a sharp intake of breath and a large exhale out the arse. He was the sweetest guy, but it reminded me why you should never get on the wrong side of him.

The next show featured Jools Holland. When I got that call I couldn't have been more delighted because I was a fan. I reckon he's the most fantastic jazz and blues pianist. We went down to his studio in Greenwich and he couldn't have been more charming and welcoming, an absolutely wonderful bloke.

He told us that when he was a kid his mum and dad used to take him to the Science Museum where he fell in love with the 'Jet 1' that's on display there. It had always been his dream to own one, and that's what he asked us to make. The problem was that Leepu doesn't do copies.

When I got back to the garage and explained what Jools wanted Leepu threw his toys out of the pram, saying, 'I am Leepu, I am the world's greatest designer! I do not copy other people's work.'

Eventually the production team took him aside and said he had to do it, it's what Jools wants, that's the deal, but he wasn't happy. It was only when we took him up to the Science Museum to see it that Leepu began to come round, when he saw that he could copy it but make subtle changes (improvements, in his mind!) while staying true to the original spirit.

My problem was that the original was based on a Rover 75 P4, a four-door cut down to two, no roof, and had a jet engine that could do 160 miles per hour. So for four grand I've got to make something that's as close to that as we are able to, and I need to make every penny stretch as far as I can.

We found an old Rover to base it on and after a few days Jools came down to have a look at what we were doing. Straight away he said it looked great, but we'd got the wrong car. We went and got a picture of the Jet 1 and, sure enough, he was right, we'd bought the wrong car! We'd bought a slightly earlier 75 and the Jet 1 was a later one, so around the headlights it was totally different.

So off we go again to find another car and dig deeper into our budget. When we find it the first thing Leepu wants to do is take the roof off, but he had been warned that if you take the roof off a four-door car it loses all its strength and bends in the middle. But with his mantra, 'I am Leepu!' off he goes and does it anyway, and sure enough it bends in the centre, so we had to strengthen the entire car, meaning more work and more delays.

The original engine was like a hairdryer, it wouldn't blow

the skin off a fucking rice pudding, and somehow I've got to turn this into a supercar. I'm racking my brains trying to think of what British-made engine could give us that sort of power without being too expensive.

Then in the middle of the night, the time when all the best ideas come to you, I got it. I went in the next day and told the production team I needed an old Daimler, 6-cylinder, 4.2 litre engine. So that's what we went out and found an old Daimler – that's basically a Jag with a bit of trim. I took out the engine, gearbox and back end, all the wiring loom, suspension and front beam, threw the body away and transplanted what was left into a chassis that was fifty years older.

You can imagine the amount of engineering required to make it all fit. I stripped and built the engine, changed all the carburation to Webber carburettors, up-ratioed the gearbox, remade all the wiring loom, exhaust and everything. Then we go to put the body back on the chassis.

And it don't fit!

We're about two inches out, so we do a bit more cutting and chopping and eventually it fits. We put everything together, started her up, and it turned over like a dream and looked the spitting image of the Jet 1, with just a slight modification on the wing.

Next we have to road test it. We'd taken Jools to meet the original designer and he told us it wouldn't work, it would be totally unstable. Our car was only tack-welded, which means that instead of continuous welding along the joints it was done about every inch. We took it up to Bovingdon Airfield and drove it at 148 miles an hour! It had no windscreen and so muggins here had to put on his Biggles flying goggles and drive this thing at twice the national speed limit, knowing that it was only tack-welded together – and I only had a lap belt on! Thankfully it was completely stable, so much so I could take my hands off the steering wheel. It went like shit off a shovel.

The Angry Frog is one I'm very proud of, but Jet 1 is the one I am absolutely over the moon with. It was one of my proudest moments as a mechanic. I think we almost doubled the budget on that car but it was worth it for the result. We went to the Albert Hall where Jools was performing that evening to hand over the keys and you could've literally knocked him down with a feather. We'd sourced dials and gauges from an old Spitfire to use in the dashboard, sprayed it racing green and done all the interior out in red leather. It looked stunning and you could see there was a tear in Jools' eye: we'd realised his dream.

The other stars we made cars for in the series were England rugby captain Lawrence Dallaglio, French footballer David Ginola and comedian Johnny Vegas. I just have to mention Johnny because I don't think I've met anyone who made me laugh as much as that guy. He was exactly as you see him on TV: always laughing, always joking, drunk as a bloody skunk, but he had us in hysterics the whole time.

Once we'd finished the series, that seemed to be it; the *Chop Shop* closed its doors for the final time. I think we'd achieved all we could and also, with that final series, the celebrities were very well known in this country but I'm not sure how many people in Finland or Turkey cared whether Martin Kemp got the car of his dreams or not, so perhaps it didn't rate as well as the others. But man – what an experience. Two years of my life that I will never forget, and I couldn't have done it without that mad, fat Bangladeshi bastard Leepu.

Off set, away from work, Leepu was a really dedicated family man and a great laugh, he would always have me in stitches. But above all else he's a metal freak. He is never happier than when he is working with metal, shaping it or crafting it in some way. Where my wife is a Shopaholic, Leepu is a Chopaholic. However, he likes to cut corners in every sense. He was always so eager to get things done, it could never be done quick enough for him. The problem was – though it made great TV – Leepu would always be so focused on what he wanted to do, he never took advice or considered the consequences of his actions. He's not a mechanic, he's a designer, so all he was concerned about was what a car looked like and how he would achieve it.

When he got a project, he would go into this strange meditation mode, sometimes helped along by four or five large whiskies, and the design would slowly form in his mind. Once it had, he would be like a Whirling Dervish; he'd be like, 'Right, I want to chop the body.'

Trouble is, you can't just chop a car because you've got all the flammables in it – you've got the battery containing sulphuric acid, you've got oils in the engine, and fuel pipes. So before he starts chopping it, what we had to do was prepare the car properly.

Sometimes he'd come in very early in the morning without telling anyone. He was so eager to get going, that he'd start chopping without draining any of the flammables. I mean, that's just suicidal – he could go up in a fireball, or blow himself to pieces! Once he got the design in his head he would never waver. Nothing to do with the safety of the car or anything else mattered to him, it was just the lines that he cared about. He would never use rulers, all the sketching was freehand on paper and then he just cut the metal with an angle grinder. In his own mad way he was a genius. The only problem was that he drove everyone else mad too.

He wants to be the world's greatest designer and to him everyone else was just there to interfere with what he was doing. And yes, at times I was interfering, because the car needed to be safe, but Leepu didn't see it that way. As far as he is concerned, if it looks right it is right.

The thing is, in the West, we are weighed down by health and safety – as a mechanic my first priority is the safety of the vehicle – but where Leepu comes from, health and safety is the least of people's worries. Life there is often short anyway; no one has any money, so they can't afford to be safety-conscious. They just need a car that gets them about, any which way they can do it. Any time you modify or change a car over here you have to pass a thing called the SVA, the Single Vehicle Approval, whereas over in Bangladesh you can start off with a bus and end up with a tricycle. No one cares as long as it goes. All I would ever get from him is, 'This is bullshit! I am Leepu, I am the world's greatest designer, this is what we must do.'

I suppose in a way we were two sides of the same coin. While I didn't always agree with Leepu's methods, I saw in him the same focus, dedication and unwavering determination to get the job done that I hope others see in me. For that reason there was a very deep respect between us, so despite the arguments, we saw how the other's work improved upon what we had done and that, in the end, is what made us a great team.

Nowadays we still email occasionally, but the longer we go without seeing each other the more I realise we are from different worlds. He's now in America, so he's living a very different life. I knew the TV stuff wasn't going to last so I went back to my day job as soon as it was done.

When I found out I was going to be doing another TV

series, *Classic Car Rescue*, I emailed him to tell him my good news, but never heard back and that really hurt me.

This hurt me because we used to be such friends. When he was in London and he brought his family over to an unfamiliar country, his wife and my Lisa would go out together, we looked after them. But I can't say we are close now. Last I saw was a picture of him on Facebook working from what looked like a garage at the back of a house. I don't know if life has been good to him or not recently.

I hope it has.

CHAPTER TWELVE

FUCK ME I'M FAMOUS!

Because the Discovery Channel is in almost every country in the world, it seemed that I was known the world over and soon offers were coming in from all sorts of places. One of the most memorable was from the *Castrol Extreme* show in South Africa. They asked if I'd like to be their headline guest and for this they'd fly me out to Johannesburg, put me up in a fancy hotel, pay for all my food plus £850 a day for ten days. I have to admit I was struggling to find a reason to say no!

I had no idea how big a deal this event was until I got out there – the main auditorium alone held 3,000 people. Four times a day I had to go on and talk for twenty-five minutes in front of all these people – but I'm no showman, so what the bloody hell am I meant to do? I had no idea if anyone knew who I was or what I was doing there, so when I came out for the first time I was relieved and stunned to hear a huge roar from the crowd. People were chanting my name and asking where Leepu was, it was crazy! I had nothing prepared so I just blagged my way through it.

So I asked who had seen *Chop Shop* and about 80 per cent of the audience put their hands up. I then asked who hadn't and a few brave souls put their hands up too, so I said 'You haven't seen *Chop Shop*? Well there's the door, now fuck off!' This got another huge roar from the crowd and from then on they were on my side. I'd tell a few stories, answer a few questions, but I realised after a while it didn't matter what I did – they were all petrol heads and they were just pleased to see me. The event normally finished at 6.30 pm but sometimes I was still there at nine o'clock at night still signing autographs and chatting to people. The organisers were so pleased they invited me back three years running.

After the third year, Graham, the organiser of the show, called me and said he was friendly with the owner of a magazine out there called *Modified* and would I be interested in writing some articles for them. Me, with my literacy skills? I ask you! All right, I said, I'll give it a go. So in my spare time I started writing about various aspects of mechanical work, modifying cars plus a few choice stories from my career.

As the months went on people would send in questions to the magazine for me to answer and my contributions grew and grew, until eventually they made me Editor-in-Chief! This I was still able to do from the UK, but after a while they called me and said they'd sorted out a regular spot on the local radio station in Durban and they wanted me on it. It was just a week so I said I was happy to.

They put me up in a hotel for the week, I met the team from the magazine, did my radio spots as the 'Car Doctor' and, just as I was due to leave, the owner of the publisher asked if I could stay a little longer. He felt that to really get the magazine going it would be good for me to stay for a few months and work with the team. They were paying me nicely and it was another little adventure so I agreed.

All was going well until two months into my stay, when something happened that changed my life, though I've never told a living soul about it before now. This is going to be news to everyone who knows me, even my wife. Despite the great strides South Africa has made since the days of apartheid it is still an incredibly dangerous country, particularly Durban where I was staying.

I was told never to drive after dark, but if you have to do so, then don't stop at traffic lights if you can help it, lock your doors and keep your windows wound up. One night I was driving home after seeing some friends and even though it is 11 o'clock at night it's still sweltering, but I can't be doing with air con, it dries my throat, so I always drive with the windows down.

So I pull up to a set of traffic lights and just as I'm checking to see if the coast is clear, I feel something pressed to my temple. It's cold, hard and I know immediately what it is. I keep facing straight ahead but out of the corner of my eye I can see it's a young black guy, skinny as a runt, with wild eyes. He's clearly high on something. He starts screaming at me: 'GIVE ME YOUR MONEY MOTHERFUCKER, I'LL FUCKING KILL YOU, GIVE ME YOUR MONEY!'

I realised it was fight or flight. Out there they'll take your money and still kill you, so I had to do something and do it quick. 'OK OK!' I shout, 'Take it easy, you can have my money.' I move to get my wallet out of my pocket and then, without really thinking what I was doing, as if I'm on autopilot, I just lean right out the car and punch him harder than I've ever punched anyone before. This gave me the split second I needed to put my foot down and speed off as fast as I can.

Quarter of a mile down the road I turn off, make sure I'm out of sight, and pull over. I scramble out of the car. My heart is beating, I'm dazed, and I'm sick right there by the side of the road. After a few seconds of catching my breath, I got back in the car and drove straight back to my hotel, threw my belongings into my suitcase and headed straight for the airport. I was on a flight home the following morning.

I couldn't tell Lisa what had happened – she'd have belted me for being so stupid, but would also have been worried sick about what could've happened. Even though I was now safe, she'd still be having nightmares that I was killed. I didn't want her to worry about me going away in the future and, to be honest, I didn't want to think about it myself. Talking about it would've brought it all back, so when I got home I just said I'd wanted to surprise her and was always planning to come home after two months. The longer I left it the more difficult it would've been to come clean, so I never told anyone.

Until now.

After *Chop Shop* finished, and in between my stints as a South African celebrity and motoring journalist, I went back to the day job. But I'd got the bug. The three series I'd made for Discovery were so well received, I was getting Facebook requests and recognised in the street all the time. I knew people liked what I did and so I never thought for a second it would be over. So I decided that if I was going to do any more TV I needed to do it properly, and I decided to get an agent.

I had no idea how TV was made, no idea what it costs or how much I should be paid. Raw Television had made me an offer, it was a decent wage and an experience so I thought I'd go for it, I had nothing to lose. But making *Chop Shop* made me realise I had a lot to learn when it came to the world of TV and I needed help.

With a view to getting an agent we set off on rounds of meetings. You have no idea how many different production companies there are until you are in the business. It seems there's one on every street corner. Everyone is very friendly and enthusiastic and assures me that they can get me back on TV. But it isn't just a matter of having a great character and a great idea; in the world of television there are millions of variables that make the difference between a show being commissioned and one that ends up in the bin.

During this time there are two little projects I did that I'm incredibly proud of, even if they're not the things that most people remember me for. The first, in 2009, was a show called *Young Mechanic of the Year* for BBC Three. It was hosted by George Lamb, son of the actor Larry Lamb from *EastEnders* and *Gavin & Stacey* and I was a judge alongside a guy called Dave Massey.

As the title suggests they'd assembled the best young mechanics in the country and we were to put them through their paces in order to crown a champion. As you can imagine, my mind went back to some fifty years previously when I was sitting my mechanic exams as a teenager and those practical tasks they made us do. It was very satisfying to turn the tables and suddenly be the one in charge of setting the faults, holding the stopwatch and checking their work.

And I really was so pleased with the standard of contestants. I love working with young people who are eager to learn and the talent on display was fantastic. It warms an old fart's heart to know that his industry is in good hands after I go to the Great Garage in the Sky.

The next programme I did was a series of short films for

SuperScrimpers on Channel 4. This was just after the financial crash and for a lot of people money was tight, as it still is, so the show gave viewers clever little tips on how to save money on all sorts of things. It was mostly old ladies who grew up in the time of rationing after the war and I suppose it was similar for me. We live in a throwaway culture these days: if it's broken just chuck it and get a new one, but when I started out in garages you made do and mended. So the production company, Endemol, were delighted when I gave them a long list of things people could do to help save them money on their motoring. You can find some of these in Chapter Sixteen of this book.

After I'd recorded a few short films with them they asked if I would run a masterclass. They invited a group of young women down to the garage for me to give them an introductory lesson in the basics of car maintenance. These girls were all in their teens and twenties and some of them didn't even know how to open their bonnet, let alone ever having looked under it before.

But I was amazed at how quickly they picked things up. I had them replacing radiator filters, changing the oil, checking tyre pressure... everything you need to know to keep your car safely on the road and avoid unnecessary bills at the garage. If these girls could do it then anyone could. It made me realise that there's simply no excuse for anyone not to know this stuff. So if you bring your motor into my garage with a clogged radiator filter, you'll get no sympathy from me!

Then I got the call I'd been hoping for. Channel 5 were making a new series about restoring classic cars and they wanted me, once again, to be the man to make these old bangers run. And once again they wanted to pair me with someone who made me look slim! After two years with that fat Bangladeshi Leepu, I was now going to be working with a fat Canadian-Italian called Mario Paceone, who also had an attitude problem! I don't half fucking find 'em, I tell ya.

Mario is a lovely, lovely guy, but what he knows about mechanical work you could probably write on the back of a stamp. He loves to wind people up, but he was a really charming guy, though, and could use his charm to wheel and deal like no one I've met before or since. His motto was, 'I never pay what anyone ever asks.' He loved a bargain, old Mario, and he usually got it.

The first time we were introduced was when they filmed a 'taster tape' to see what we were like on screen together. He came up to me and said, 'Hello Baldy.'

'Fat cunt,' I said in reply, with a smile. And it all went downhill from there! We were always having digs at each other, but it was all in good spirits. I knew from that first exchange we would get on – we could give and take it and were always making each other laugh. I don't think I've ever laughed so much in my life as when one of the doors jammed on one of the cars and Mario had to climb in through the window. He got wedged in at the hips, feet waving in the air, and it looked like the car was giving birth to a fat, hairy baby!

He was another one who enjoyed the London life and went home about 20 lbs heavier than he arrived. All he did was eat and shit, he certainly never did any bloody work. And as if to prove the point, every day he would go to have his nails manicured. Fucking manicured! One that's never done a day's work in his life, that's who. His hands were softer than the cheeks on my arse.

By this time I knew that in TV there is never enough time or money to do what you want to do, and once again the budgets and timeframe to get the work done were unrealistic, we all knew that, but as ever we worked our arses off and got some pretty extraordinary results in the process. Of course, working in such close proximity meant that me and Mario had our ups and downs, but we always came out smiling in the end.

The series was a co-production between Channel 5 in the UK and Discovery Canada. In the first series we made four cars in Canon's Park, Edgware, Middlesex and two more in Toronto, Canada. So we were doing twice as many cars in the UK, but had the same amount of time in both places to do the cars. It seems a different pace of life over in Canada. Whereas here we would be working until two, three in the morning to get things done, over there they started at 9.30 am and at 5.30 pm on the dot they would say, 'That's it, we're off.'

The most memorable one from the first series was the Mini, which was as rotten as a bloody pear when we bought it. So the first thing we had to do was make the car safe, so we proceeded to cut all the old panels out, put new panels in, then needed to check that the car was straight after all the chopping and banging that had been done to it. So we call these recovery vehicle guys to get it taken over to where it was going to be tested. These three herberts turn up, load it onto the back of the truck, then say, 'Right, we want paying now.' Fat chance! I want to make sure it gets there in one piece first. They're having none of it, and I'm having none of it!

'Look,' I said, 'this garage has been here thirty bloody years, we're not going anywhere, I want to know the car gets there before you get your money.' They said they didn't work like that so I told them to take the car off their vehicle and said we'd call a proper company instead. They're fuming by this stage, saying I'm wasting their time, so rather than drop the back of the truck to wheel it off they pushed it straight off the back, straight onto its fucking roof! I chased them out of there sharpish, I can tell you!

The VW Camper we built in Canada is one of my favourite cars I've ever worked on. I'd go as far as to say it was the best Camper I've ever seen, even if I'm being slightly biased. I was working with two guys Diego and Satnam in the Toronto garage and it had some of the finest workmanship I've come across. It was finished superbly, in silver and blue, with lots of chrome and beautiful curves – and it even had its own fold-out barbecue in the back! I must say, it looked a million Canadian dollars when we'd finished.

At the other end of the size spectrum the little Fiat 500 we worked on in the UK was a beautiful little motor, the attention to detail was amazing, and restoring that to its former glory and then some was a pleasure. I mean, it still only did 0–100 mph in 5 minutes 10 seconds. Sorry, I exaggerate – 25 minutes and 10 seconds – but it really was a lovely little car.

In contrast, the E-Type we bought was one of the worst examples of the E-Type I've ever come across, and just remember I've been working on these things virtually since the day they came out, and I've seen some shit in my time. This one, though, was the pits. Left-hand drive, sunroof, automatic... Everything you don't want on an E-Type, this one had, and it still cost us £13,000! It was the cheapest we could find so we had to make do with what we could afford. Any decent E-Type will cost you a minimum of twenty grand, if you're lucky. Meanwhile we had to make a silk purse out of a sow's ear. We did it, I think, but only just.

My favourite from Series One was the MGB – so much so I tried to win it! The idea of the show was that at the end the viewers had the chance to win the car we'd made in that episode. You dial a premium rate number and you go into the hat. Well, I really fancied that MGB so I called up about ten times, it cost me a bloody fortune. Then my agent pointed out to me that under the terms and conditions, no one connected with the programme was allowed to enter, so I'd wasted my money.

He could've told me that earlier, clever-clogs!

CHAPTER THIRTEEN
THE SAMARITANS

My father was ill. I'd just dropped him off at home after coming back from the hospital where he'd told the specialist in no uncertain terms that he was not going to have his testicles removed.

I sat outside the house in my car, and I felt completely lost. I knew in my heart I was losing my dad. I didn't want to burden my family, I didn't want to burden anyone with it, but I was a wreck, I didn't know how I was going to cope without my dad. For once in my life I didn't know what to say and didn't know what to do. I just sat in my car for hours and hours and cried my eyes out.

Eventually I drove to the shops, got myself something to eat and picked up a paper. In the paper there was an advertisement for the Samaritans that said: 'Call us if you feel lonely, despairing or suicidal.'

I thought, well I'm lonely and despairing, two out of three, that's me. I had a car phone then, and so I sat in the car and dialled the number. I didn't really know what I was going to say, I had no idea what I was doing, but found myself calling them. The person at the other end of the line said, 'The Samaritans, can I help you?'

I couldn't talk for a minute. I wanted to speak but couldn't. The person said into the silence, 'Take your time, talk to me whenever you're ready, I'm here to help you.'

For five minutes, I said nothing.

'Can I ask your name?' the Samaritan asked.

'Bernie.'

'Bernie, what's troubling you?'

Then all of a sudden it was like a dam breaking, it all came out. I was on the phone for two, two-and-a-half hours, and told them everything. I couldn't stop talking, couldn't stop crying.

Afterwards, I just felt better. I can't describe the feeling, but I felt better. I knew I had someone I could talk to, someone who couldn't see me, knew nothing about me, wouldn't judge me, someone who would *just listen*.

For the next few months while my dad slowly passed away I called them two or three times a week, sometimes very late at night, and they helped me through it. When my dad finally died I was able to accept it and I don't think I'd have been able to, had I not been able to get it all off my chest during that period. I promised myself then that once I'd sorted myself out, I wanted to give back what was given to me.

A few weeks later I applied to become a Samaritan. I had no idea what the training would be like, and it was incredibly in-depth. You had a Samaritan psychologist assess you and they wanted to pull everything out of you that might be troubling you. They wanted to see if you were depressed. They wanted to know if you had any prejudices. They delved into my background, my private life, made sure I had no criminal record. I went back three or four times and once they were satisfied they'd heard everything they did an evaluation and I passed, though every time my dad was mentioned I cried, I couldn't help it. I still cry now.

The training was intense too. You had to learn how to listen without talking and also just how to listen. There were certain things you had to listen out for, like 'pick-up points'. A pick-up point is when someone says something but then immediately dismisses it or changes the subject. For instance, they might say, 'This happened, but it didn't affect me.' So rather than letting that point drop, you would say, 'OK, how do you feel about that now?' Often you would say that and they would realise, or they would admit to themselves, that this issue was part of their problem, or it would lead onto something else that was the problem. You were opening doors all the time.

There are all sorts of protocols. The Samaritans were founded by Reverend Chad Varah, and although it was run by the church, you never brought any form of religion into it. You never said, 'Pray to God and it will all be better', or anything like that. You never give your real name, you always choose a pseudonym, and you never speak to the same person twice. Fair enough, you might if they call back and you happen to pick up the phone, but if they make the call and ask for you again you say, 'I'm sorry, Terry or whoever is on another call at the moment but I can speak to you.'

You log every call with the name of the caller, your name, the reason they called, how long the call lasted and whether they want a follow-up call. This follow-up, again, would be done by someone else. It is a safeguarding measure, making sure you don't get too emotionally involved in someone's problems.

And you never give advice, but they have all sorts of leaflets and phone numbers people can call if they have particular problems. Maybe they have housing issues or something, so you can say, 'If you like, you can call this number and they will be able to help you.' It's not giving advice, just giving information which may be useful to them for their specific problem.

It is up to the Samaritan who takes the call to judge if a caller is suicidal, or a suicide risk. On every watch there is a crash team of two or three people who go out to be with people if they are thinking of committing suicide and want someone with them. Sometimes they may want you just to stay on the phone with them while they do it, which you would, but sometimes they don't want to be alone and so they ask for someone to be with them.

However you're not allowed to talk anyone out of committing suicide. By Samaritan Law a Samaritan is not allowed to call an ambulance if they have taken an overdose or whatever. You say to them, 'If you want me to stay with you then I will do, but I cannot call an ambulance for you. But, by the law of the land, if you lapse into unconsciousness then at that point I have to call an ambulance.' Several times I was on the crash team and went to people's houses and sat with them while they took an overdose, though more often than not they would just talk about it.

I remember one Bank Holiday Monday me and Lisa were meant to go out, but at eight o'clock in the morning I got a phone call from the Samaritans saying they were short-staffed and could I go in to help them out. I say 'could' but I was a Samaritan Leader and so it was my duty to go in, so that was our Bank Holiday screwed.

Not only that but we only had one car at the time which I had to take to get into the Samaritans, so Lisa was stuck at home. After an hour or so we get a call in from a woman who says she can't take it anymore, her husband has left her and there's nothing worth living for, she's taken a whole bottle of diazepam and she wants to die. But she wants someone with her, and so me and the other member of the crash team jump in my car and we drive round to her house.

When we arrive she shows us the empty bottle and tells us how her husband has left her after over twenty years of marriage, that she doesn't have any children and has nothing to live for. She's clearly taken an overdose but we can't call an ambulance until she loses consciousness. All of a sudden her eyes start to roll and she begins to fall asleep. We begin frantically calling 999 but we can't get an ambulance. We're waiting and waiting but nothing is coming, so we pick her up, put her in the back of the car and I race us all up to Watford General Hospital. We pull up outside A&E and they take her straight inside, pump her stomach and save her life.

Unfortunately, on the way there, she'd puked in the back of my car, all over the seats, all over the floor, and all over my colleague who was sitting with her in the back! Now, not only have I left Lisa on a Bank Holiday, who I'd promised to take out when I returned, but now our only car is now sprayed from floor to ceiling in vomit.

I spent the rest of the day scrubbing all this sick off, putting lemon juice on it to neutralise the smell and cleaning and cleaning until I'd got it all off. When I got home my wife was not happy, I assure you! But, at the end of the day, she couldn't be mad with me for too long. I had probably saved this woman's life, though I will never know for sure. As a

Samaritan we don't stay with them at the hospital – we can't get involved in that way, we just do our bit and leave the rest to the professionals.

Twice a week, I would do a four-hour shift, then once a month I'd do a full night duty. I also used to do one-on-one interviews with people who came to Samaritans House in Watford. You ask them if they'd like a cup of tea, but otherwise it is like a phone call; you try not to talk, just listen to what they have to say. It's a bit like counselling. But you have to be very careful – you park your car out of sight and enter via the back door, trying not to be seen. You have to be aware of anyone loitering around, in case they are waiting for a particular volunteer, and you have to be careful when you come back out, get to your car and drive off without anyone following you. You have to be as anonymous as possible.

I also enjoyed going round with the collection tin. Some people would just stand outside a Sainsbury's somewhere rattling the box, but I'd go up to people who were trying to ignore me and say, 'Come on you mean bastard, it's a good cause, put some money in!' Always with a smile on my face, of course, and it usually worked. Sometimes people would say they didn't have any change, so I'd make a note of their face and when they came out after their shopping I made sure I caught them then instead!

In all I was a Samaritan for nineteen years and it is one of the most rewarding things I've done in my life. Unfortunately a few years back I could no longer commit to doing the night duties and you can't be a Samaritan if you can't do night duties, it's not fair on the rest of the volunteers. Also, as I was on the television and getting recognised in the street, I didn't want to be in the awkward situation of someone coming in for a one-to-one and potentially recognising me.

The point of the service is that it's anonymous and you don't know the person you are talking to, so if I was recognised that would change the dynamic of the conversation and not help that person. I sincerely hope, though, that in the future when I have a bit more free time that I am able to go back. They're an amazing bunch of people who care so deeply for others. It's an incredible organisation and it's a privilege to help people in need.

I've certainly never forgotten how they helped me.

CHAPTER FOURTEEN

GORDON RAMSAY, EAT YOUR HEART OUT!

In between the various television appearances I was still doing the day job. I had been in the business for over forty years by now, worked at a lot of places and my name had travelled to many more. From my consultancy work to my intervention at the garage where I found the receptionist, Paula, with her hand in the till, I had always had a reputation for fixing problems with companies. With my new-found status as a TV personality I started getting calls from garages that had issues and I was asked to go in and give my professional opinion.

And so I started to become the 'Gordon Ramsay of Garages' if you like, going into businesses, seeing what was wrong, hiring and firing and turning the place around. In this job the biggest problem I find is getting people to open up and admit they have a problem. Even as a Samaritan it's not always easy in the mucho macho world of the motor trade to get guys to say, 'Please can you help me?' You can't help someone who doesn't want any help or doesn't realise they need it.

Once they've admitted they have a problem, it's then a case of working out what that problem is.

In many of the garages that I have gone to where I have been recommended by friends or colleagues, they say, 'Business here isn't good.' So when I go in, the first thing I look for is whether they have returning customers, and if not why not. Is it the fault of the mechanics? The management? The customer service? The pricing? You have to go into all of these things to find out exactly what is causing the business to go down.

I remember one particular garage in Finchley. The place specialised in one make of car, which was Alfa Romeo, and a very good friend of mine was also friends with the owner and used to take his car there. But he'd say to me, 'You should have a look at this garage, I don't understand it, whenever I go there it's dead, they're doing terrible.'

This I thought was strange. OK, Alphas are a niche market but they are well known for having 'teething problems', shall we say? (That's my polite way of saying they fall to bits on a regular basis.) So I would've thought a garage that specialised in Alphas would have a steady stream of business. My mate said he knew the owner and said he was a really lovely guy, he needed a bit of help and suggested that he could benefit from my expertise and experience.

So I gave the bloke a ring and, sure enough, he's a really nice guy. Not a mechanic himself, he's a businessman, and he's had the garage for about twelve years. I tell him all about myself and my experience and so he invites me in for a cup of coffee and a nosey around. I go in and it's quite a clean garage, nicely laid out, but all the mechanics are sat around doing nothing. They had four mechanics and a receptionist and only one car on the go, nowhere near enough work to keep all of these people employed. The boss can't understand it and it doesn't seem right to me either, as I know they do a decent job on my friend's car, so I offer to come and work for them for a week and see what I can do.

I start the following Monday and the first thing I do is go and meet the receptionist and look through the books. I asked him why he thought they didn't get returning clients. He said they'd had quite a few complaints about dents in cars, and things going missing from the vehicle interiors. I asked if they'd broached this subject with the mechanics. He said he had but no one ever admitted to anything, they all covered each other's backs, and all they were interested in was putting their hand out on a Friday afternoon and getting paid. They all worked on a flat rate, so whether they do twenty cars or one, they all get good wages and they don't give a fuck.

Next I asked to be introduced to the guys, so we called a meeting in the tearoom. They were naturally sceptical of me, and were either stand-offish or downright rude. I went over my background with them, told them about Rolls-Royce, Jaguar etc.

'Yeah, but that's Rolls-Royce, Rolls ain't Alphas, what do you know about Alphas?' one of them argued with me.

'I know enough,' I say.

So they fire off lots of questions, which I answer.

After about an hour-and-a-half of this I had a pretty good overview of what they were like. I went outside and saw the receptionist and said, 'You need to call the owner, I need to have a chat with him.'

The owner came down and I said, 'Look, there's only one way you're going to be able to turn this place around. You've got six mechanics altogether, you've got to get rid of four of them because they're useless. They don't give a damn about what they do, at least one person amongst them is stealing from cars – and when this happens those clients don't come back, so they're effectively stealing money from your pocket.'

At least four of them had to go, but I couldn't make them redundant, so I had to play it a little bit clever. We called a meeting with the whole firm and the owner said that he was handing over the running of the business to me and that I'd have as much authority over everyone as he does.

As soon as he'd finished I went in all guns blazing, starting with: 'I've had an hour-and–a-half with you lot, I've got a pretty good idea of what you're about and I have to say I'm not happy.'

'What do you mean you're not happy?' someone called out.

'I mean stuff is being stolen, cars are being returned dented, no one is taking responsibility, so I'm calling the police. So, for whoever is nicking stuff from the cars I have some advice: go and collect your tools and get out now, because as soon as the police are involved you're on your own and you are going to be in serious trouble.'

With that four of them got up and headed for the door. I told them as they left, 'Good, because if you hadn't walked out, you'd have been fired anyway.'

I had no intention of calling the police, but this saved a lot of hassle.

So now I'm down to two mechanics, the two that really wanted to be there. We all set about calling former clients and asking them to bring their cars back, or we would go and collect them to save them the trouble, and along with the two decent mechanics we started going over the previous work that had been done, all for free.

This gave the company good grace, and slowly but surely the clients started to come back. We put a big advert in the local paper inviting anyone with a previous grievance with the garage to get in touch. They got calls and they brought the cars in and put the faults right, for free.

I was with the company, off and on, for three-and-a-half months and in this time it went from a dead garage losing tens of thousands of pounds a month to the busy, bustling and successful business I thought it should be. It went from being on death's door to rude health and went on for another ten or twelve years before the owner sold out for a very healthy profit.

That garage was a prime example of how just one or two rogue employees can ruin a reputation. Word of mouth, particularly in a clique like Alpha owners, gets bad news around quickly and overnight you can lose business. This success, along with many others, means I'm now in high demand as a consultant for various garages in the south-east of England.

When I look back over my career I've done things I'd never expected to do. I had no idea back when I was nine years old and I was taking my uncle's new vacuum cleaner apart, or walking into that greasy garage when I was twelve, that it would lead me to working in far flung corners of the globe. Certainly, when I was working for Al and the other characters at the Blind Beggar, I had no idea I would end up working for, and becoming indispensable to, the Metropolitan Police.

On and off for the past few years I've worked directly and indirectly for the Met. I started off simply driving recovery vehicles. I was happy to get out of the garage once in a while so I was sent to police stations all around London and as far afield as Gravesend in Kent, collecting vehicles that needed work doing and bringing them back to the main workshop in Park Royal.

It seemed that every policeman in London watched *Chop*

Shop, and every police station I went into I'd get: 'Fuck me it's Bernie the Bolt! Where's Leepu?' Then I began working on the cars themselves, sometimes on the spot at the station or back at the workshop, and of course attending breakdowns out on the street. More often than not, this was because diesel had been put in a petrol engine, or petrol in a diesel. So yes, even the police make that mistake sometimes!

And it was really interesting work, getting to see how the police went about their business, and working on vehicles from high-performance unmarked vehicles to riot vans to those used by the Diplomatic Protection Group.

The only part that wasn't so good was attending the accidents. Unfortunately, during high-speed chases, mistakes are made or there is mechanical failure and things go wrong. Sometimes they go fatally wrong. The accidents you really don't like to see are the ones where officers have been trapped inside vehicles and had to be cut out. When an accident involving a police vehicle occurred I would be scrambled to the scene to move the vehicle as quickly as possible. I didn't touch the civilian vehicle involved (if there was one) – that wasn't my remit, I just had to get the police car back to the garage pronto. Sadly a stricken police vehicle in certain parts of London is a magnet for troublemakers and it would get smashed up pretty quickly if it were left there.

In fact the only instances where I wasn't supposed to get the car moved was if there was a fatality, in which case the crash scene investigators would have to do their thing before I was allowed to touch the vehicle. What happens is, a police car has an accident, the officer calls back to base, and they immediately get in touch with the central office of the contractors who deal with all the Met vehicles. That's the company I work for, and I'm on-call ready to go.

Patrol cars do a lot of mileage and often at very high speeds in urban areas, there's lots of revs and lots of crunching of gears, so there is a hell of a lot of work that goes on behind the scenes to keep these vehicles in the condition needed to do their job.

Getting back to the workshop was rarely a straightforward affair. I've been driving recovery trucks one way or another for forty-odd years, but nothing in my life prepared me for the first time I took a police car on the back of my recovery vehicle. One of their cars had been involved in an accident and was smashed up at the front, so I had to pick it up from Brixton Police Station. As I drove out of the yard I was subjected to the most horrific abuse: bystanders started spitting at my truck as I waited at traffic lights, and also threw fruit and all sorts. It shouldn't surprise me, I suppose, but I couldn't believe how much people hated the police. But although belligerent pedestrians were trouble, the one good thing about having a police car on the back was that no one ever took liberties with you on the road. They always gave you a wide berth and always let you out of junctions!

The worst time was during the London riots in the summer of 2011. As you can imagine there were police cars and riot vans going down all over the place, being targeted by rioters and getting smashed up. We could barely cope. I was in North London one minute, South London the next, then over to East London, you couldn't stop. I couldn't believe the damage that people were doing to these vans, smashing them up, scrawling graffiti all over them. Respect and law and order was completely out of the window.

I can remember going to a police station in south-west London and having to collect one of the area cars which was one of the big BMWs. It was a tight squeeze to reverse the truck up the side of the station, so I checked the car and it started. I parked the truck out the front, dropped the flatbed, and drove the car out by the side of the station. I was just lining it up to drive up onto the back of the truck when I was suddenly surrounded by about twenty guys kicking and rocking the car, spitting and threatening me. I thought, fuck this for a game of soldiers, whacked it into reverse and drove straight back into the station car park. I went inside and told them I was being attacked out there, so they sent out six officers in full riot gear: truncheons, shields, and everything, to protect me while I got the car on the back of the truck.

Then I jumped out and started putting the straps onto the wheels when one of the officers shouted at me, 'Forget them, just get the fuck out of here!' So I jumped in the cabin and shot off as fast as I could, with this BMW bouncing around on the back. If I'd gone over a bump it would've jumped straight off, so once I'd got what I reckoned was a safe distance away I pulled up in a quiet side street and fitted the straps. My heart was racing, I tell ya. For a few days people just weren't scared of the police anymore.

To me, it made no sense. Sure I've had my run-ins with the police over the years, but they are there to protect people and keep the peace. They are human beings just doing their job. But the mindless way these vehicles and officers were being attacked beggared belief. As well as all the damage to shops and businesses they also caused hundreds of thousands of pounds of damage, if not millions, to these police vehicles. It hurts me because having worked with the police so closely I know what good, honest people they are. In fact I am the godfather to a child of a couple who are police officers.

At my age that kind of caper is stress I can do without. God knows how the actual police officers put up with that sort of behaviour day-in day-out, so I leave the recovery vehicles to other people now.

But the Met know me very well and they trust me, and someone with an expert knowledge of motor vehicles can be useful in all sorts of ways. So once in a while I will get a phone call asking for a bit of help, but I'm afraid I can't say any more than that...

CHAPTER FIFTEEN

CLASSIC COCK-UPS

We all learn from our mistakes, but we can also learn from other people's mistakes too. As you can imagine, after fifty years working in garages I have seen some right cock-ups in my time – far too many to list here, I'd need another seven books to tell you 'em all. Like anyone I make mistakes myself, but for some reason these don't come so readily to mind, maybe I was too embarrassed about my blunders, so I blanked them out.

So here is a selection of balls-ups by both mechanics and customers that really left me scratching my head wondering what on earth they were thinking. Some are so obvious I would hope no one would ever think of repeating them anyway, but these people made mistakes and here they are, so that you don't do the same things.

In garages all mechanics think they are all full of knowledge and each thinks he's better than his colleagues. I call this the 'bullshit baffles brains' scenario. Here are some examples:

- A so-called first-class mechanic I worked with in the early days tried to find an air leak in an engine. To this day I can't fathom why he sprayed WD40 onto the manifold. As a result both he and the engine caught fire. Luckily for him another mechanic was at hand with a fire extinguisher to put it out. A lesson was learned by him, and no serious damage was done, except to his pride.

- Speaking of leaks, another mechanic tried to fix a water leak in the cooling system of a VW Beetle after the receptionist filled the worksheet out incorrectly. What he forgot was that the Beetle was an air-cooled engine, and it didn't rely on water for cooling, so he spent the best part of a morning looking for a radiator that wasn't there!

- I once worked with a mechanic called Jim the Nut, not because he was crazy, but because every bolt he fitted, he over-tightened and snapped it – every time without fail. I'm all for ensuring nuts are securely tightened but if you overdo it they break, and believe me once they've broken they are a bitch-and-a-half to get out again. It cost him so many hours in unpaid overtime trying to unthread these bolts and put new ones in that he eventually left.

- Putting a car on a vehicle lift needs some knowledge about where to put the arms of the lift. One guy was in such a hurry to lift the car he placed the arms in the wrong places: not on the correct jacking or secure points designed to take the weight. This resulted in the arms going through the car's body and a very costly repair which had to come out of his wages. And we're talking probably hundreds, if not a grand or more. What's that old saying, more haste less speed?

- Never dick about with a hot engine – they can be lethal. A couple of years ago I saw a so-called mechanic remove the water cap from a hot engine, whilst it was running. I ask you, what an idiot! The boiling water scalded his face and arms, it was terrible to see, as if it happened in slow motion, but we got him under the cold tap as quickly as we could. He got a few scars from that one, I tell you. He certainly learnt his lesson and hopefully he learnt it for your benefit too.

- Replacing an alternator is quite a simple job, providing you disconnect the battery. One fool (a mature mechanic with an attitude) did not. As he removed the alternator the main live wire touched earth, resulting in a near burnt-out car. Had the apprentice not reacted quickly and cut the live wire at the battery the car could have been toast – and the guvnor would've made that mechanic the margarine on top of it!

- My pet hate is a mechanic who thinks he knows everything. In fact, you never stop learning. One day this Mercedes A Class automatic comes into the garage in 'limp home' mode, with a warning light on the dash. Our expert diagnosed transmission problems, carried out a scan code and filled out the job sheet. The client was called, and it turned out that this expensive repair of the transmission would cost about £1,800 plus VAT. Go ahead, the client says, and the mechanic duly removes the front frame complete with the transmission. Replacement unit fitted, car started, scan coded and the same problem is there. Wonderbrain says: 'It's a faulty unit,' so he again removes the front frame and fits another one. SAME PROBLEM. I then noticed when he went to test drive the car it had no brake lights, so it was just a faulty brake switch all along. This cost the client about £8 and problem solved. The mechanic was sacked for arrogance after he verbally assaulted the manager when he was in a foul temper. Pride before a fall and all that.

• Another head-scratcher involving brake lights, this time for me. One of the secretaries' cars was a Nissan Primera. It would idle perfect, rev-up stationary perfect, but on the road there was no power. I scan coded the car, no faults, tried again and no faults, live data road test, no faults. Checked for a blocked catalytic converter, no fault. I'm racking my brains, thinking what on earth is going on? Then I noticed she had no brake lights. Nissan, in their eternal wisdom, put in a safety connector from the brake circuit to the engine ECU – so that if the brake lights don't work you are limited to 50 per cent less speed. It was a 'safety feature'. This will also apply if you fit LED bulbs, as the circuit is then not complete. Brake bulbs replaced, no further fault. I learnt something from this: NO ONE is all knowledge, not even an oldie like me.

• As you can imagine there is a lot of machismo in a garage, a lot of testosterone flying around, and occasionally you get a mechanic who likes to show just how macho he is. Well, a garage I was doing consultancy work for had one of these. This fool was refitting a Chrysler Voyager clutch when the transmission lift packed up. Instead of asking for help from another mechanic to lift in the transmission he lifted it himself, got halfway up and dropped it. The casing smashed to pieces and a new transmission had to be fitted. Our Macho Man had to pay it back weekly from his wages, and it cost him about £750.

- Lesson learnt with power tools. A mechanic replaced the track rod ends in a Mercedes CLK, re-tracked the wheels and went for a road test. The steering went, and he nearly hit a wall. This fool had used an air gun (a compressed-air driven tool that can be used as a spanner) to tighten the track rods. The ball socket literally pulled through the housing on road test. NEVER tighten these or steering components with air guns, they have to be torqued up to a specific poundage (i.e. tightened to a specific degree, using a torque wrench adjusted to the correct setting). You cannot tighten this kind of joint to the correct degree with an air-powered gun.

- George Orwell was right, Big Brother is watching you... In one place where I worked we got a basic wage plus commission on all the jobs we completed, so if you were quick you earned good money, or you'd earn from advising customers to have extra work. Now as you know, I would never deceive a customer and the dealership itself wasn't in the business of trying to scam people either, but the prat mechanic on the next bay to me pushed a screwdriver through the radiator of a car to get extra work and to sell a new radiator. The trouble was he didn't realise that over the previous weekend CCTV had been fitted in the workshop. He was caught red-handed, sacked and prosecuted for criminal damage.

- A great result! A guy I worked with in the early days thought he was top dog, and everyone else was a fool. This was until he had to change inboard brake discs and pads on a Jaguar XJS. (Inboard brakes and calipers are located near the differential instead of on the wheels, as in most cars.). We let him struggle for four hours, knowing that the job can't be done his way. Eventually I pointed out to him that the rear frame had to be dropped to gain access to the brake discs. His words? 'Fuck off, I know what I'm doing.' OK, if you say so sunshine, I thought. Come the end of the day the car is still in bits. He went home and never came back to work. Again, his pride was broken; no wonder they say it's a deadly sin.

- I've mentioned this before, but I really hate those 'fast-fit' places. A lady brings her car in with the TomTom satnav working OK, but when it's switched on, the engine misfires and eventually cuts out. These satnavs are normally powered by a lead that is plugged into the car's cigarette-lighter socket. But this unit had a wire behind the dashboard with a live connector, all neat and tidy. We had to remove part of the dash to trace the wire, and OMG some clown had jumped a live one to the nearest live terminal, which happened to be on the main computer wiring! When I asked the client she said that it was fitted by one of the fast-fit garages who supply and fit free of charge, but she paid the optional extra £80 plus VAT for a neat wire fitting. This could have caused a fire and also burnt out the ECU, so we replaced and fitted the wire correctly. This type of bodged fitting gives the garage trade a bad name.

Not all the mistakes are made by mechanics. I know they say the customer is always right, but seriously, were this lot? The stupidity of some people beggars belief at times. They pay good money for a car and then they treat it so badly they pay even more to get it fixed, or they blame the garage or someone else for the problem. Whatever it is it never seems to be their fault.

Most of the time the problem is that they don't know what they are dealing with and are too scared to ask until it is too late. Here are some prize idiots who could do with doing an apprenticeship in a garage, or in some cases having their driving licences revoked.

CLUTCH CALAMITY

Going back a few years I remember a client had an Aston Martin DB5, a beautiful car that should be treasured like a newborn baby. His wife drove this as a status symbol, but every year the clutch was gone. She always 'rode' the clutch, and the poor thing (the clutch, not wife) had to be fixed and the husband had to shell out hundreds for the repairs. The hubby asked me if I could fit a heavy duty unit, so that she could not ride the clutch any more. I got through to a very well known clutch manufacturer, who at great expense could supply a heavy-duty clutch plate and pressure plate, which was duly fitted by myself and paid for by hubby. I told the lady to be very careful, as there was now a very sharp bite on the clutch and it was necessary to release it gently. Her reply was: 'I have been driving for over thirty years, do not tell me how to drive.' Well I think you can guess what happened next. She snatches the keys, jumps in, clutch engaged and off she goes. She collided into a wall beside the garage, causing over £1,000 of damage.

Red faced, she just gave me the keys back and walked off, not saying a word or making eye contact. Boy, was I glad I wasn't in her house that evening.

PROBLEM POO

Talking of relationships on the rocks, it's amazing what tell-tale signs a car can sometimes reveal. The number of cars that are brought in for a valet that have footprints on the headlining. And there's only one thing going on that gets footprints on the ceiling! Some valeters are more discreet than others about this, and it's not always the wife's footprints either. Then there was the case of 'the smell'. A Rolls-Royce came in with a terrible smell inside the car, and after some investigation we found a used nappy full of poo tucked under the front seat. We called the owner and his wife answered, but she said her children were all grown up. The couple were in their sixties, and she couldn't fathom where the nappy could've come from. Two hours later I get a call from the hubby, who tells me he's been kicked out of house and home. Turns out he'd been having an affair with a younger woman and got her pregnant, but had recently broken up with her. Out of spite she'd hidden the nappy under the seat. Man, he was in the shit!

FIT TO BUST

An irate client complained that his suspension was very hard and the steering was too light after we serviced his car recently. We checked all the components and found no fault. Then we checked the tyre pressures. They were set at 65 ft lbs instead of 32 ft lbs! It's a wonder they hadn't exploded! He denied any knowledge of doing the tyre pressures himself, but when I checked the job sheet it stated tyres set to 32 ft lbs.

That would be a huge error for even the most inexperienced of mechanics to make, so I called the client back, his wife answered, and I told her what I'd found out. The wife said her hubby likes to tinker with the car and last week he'd used his new tyre inflator gadget. She'd dropped him right in it! When the client came in to collect the vehicle he apologised. Maybe he'll think twice about tinkering with stuff he doesn't understand in future.

OIL ON TROUBLED WATERS

Another muppet who liked to tinker was the owner of a BMW 320i that was brought in by the AA with reported overheating. We checked the water and found that vast amounts of a milky oil-and-water mix was in the header tank, so we put a dipstick into the oil and found the same watery milky mixture there too. All these symptoms point to a blown head gasket, which can be serious, but on further chatting to the client, he mentioned he topped up all the levels himself. It seems that this plonker filled the engine with water and the header tank with oil, the doughnut! A full flush out of water and engine was carried out, the sump was removed, and a new gasket, refill with new oil, filter, water and anti-freeze done the job. The client was embarrassed, a few quid lighter in the pocket, but happy.

THE TAPPET TINKERER

Please, if you do fiddle with your car and you make a cock-up, just tell the garage from the start. You'll always get found out in the end and it will save you a lot of money for the time spent by the garage trying to find the fault. An MGB GT 1974 was booked in with terrible shaking on idle – you could feel it through the whole body. The engine mounts checked out OK, the exhaust pipes were not touching the body's underside, and the performance was low. I asked the client if any previous work been done to cause this, and he said it was perfect until the last weekend. As a precaution I removed the rocker box and, lo and behold, you could see that there were no tappet clearances, so no wonder the car was running rough with no power (the tappets open and close the valves that admit and release combustible and spent gases in sequence, so their correct adjustment is crucial to the engine's performance). I adjusted the clearances and it ran perfect. Seems the client was too embarrassed to tell me he attempted to adjust these himself with little knowledge, and ballsed it up. Fess up, it's the easiest way!

WIRE WOOL WALLY

A Porsche Cayenne 2009 comes in for a paint estimate. Three panels were literally scraped back to bare metal, and the thing was a fucking mess. I asked the client if it had it been vandalised. No, she replied, it was parked under a tree, loads of bird droppings were on it and so she thought to she'd do her hubby a good deed and clean it off with the old wire wool. Fucking WIRE WOOL! She'd caused carnage! Three panels prepared and re-sprayed at a cost of £1,000 plus VAT – probably the most expensive scrubbing pad ever!

A POSITIVELY NEGATIVE EXPERIENCE

Old Riley Pathfinder verses Triumph Stag! Our clever-dick client decided to jump-start his Riley (i.e. use the battery power from a second car – his Triumph Stag – to start his Riley) in the garage at home. His jump-leads (used to connect the cars, battery-to battery) were short so he had to put the cars bumper-to-bumper in order to attach the leads. He went to start the cars and burnt out both of them – the wiring looms and batteries were blown. Why? Well the Riley is dynamo powered, with a positive earth, and the Stag is alternator powered, with a negative earth. The bumpers touching connected the Stag's negative terminal with the Riley's positive terminal, creating 'reverse ground and live', with the result being a total short-circuit. Fifteen hundred pounds later there were no more problems, except a deep empty pocket and a client with a more comprehensive understanding of electricity.

THE BRAKE PAD BERK

A Mercedes CLK 270 came in for a brake check, and the client says to call him with a quote. Not a problem sir, I told him. The brake discs are grooved and the pads so worn that they're contacting the discs, there's no handbrake either, and the warning lights are on. A quote for repairs was given, but Mr Penny Pincher says he can get the work done down the road cheaper. I do not, and never will, back down on price, I'm not in the bartering game, but will always give loyal clients a discount. So he says, 'Thanks I'll collect my car later and take it away.' Oh no, I told him, not without a transporter, as I know that these brakes are lethal. 'No,' he insists, 'I will drive it.' Now the car is in my garage. If he has an accident it looks very bad on me to allow a car on the road in this dangerous condition, I'd be as culpable as him if anything happened. So, I have to protect myself and other road users and inform him that I will have to call the local police about the condition of the car. OK, he says. I do this, he collects the car and drives off, then gets stopped by the police just opposite the garage, who impound the car under the Dangerous Vehicle Road Traffic Act. It ended up costing him double plus recovery and I refused to have the car in the garage for the repairs. I am no snitch, but it could be my family driving in front of him, when he has no brakes on his car.

A FACE FULL OF AIR BAG

Never treat a car roughly, for sometimes they get their own back. We had a client with a Mini Cooper, and he said the horn worked only intermittently. On the steering wheel are the horn contacts and the air bag. When we stripped it down, the fault was found to be the internal contacts. I gave him a quote for a new air bag and contacts plus fitting, but the job was declined and the car collected. The next morning the same car is brought back on an AA truck. This clown had been trying to hoot another motorist, and out of frustration when it didn't work, he bashed the centre of the steering wheel so hard it activated the air bag! He got a face full of air bag, a fright, and a bigger bill for a replacement and initialisation of the new unit.

STEAM-CLEAN CATASTROPHE

A customer comes in with his pride and joy, his Audi A8. It's immaculate, we service the car and he requests an engine steam clean. Now, with all the modern day electronics, the worst thing you can do is get water in the electronics, so the client takes the car after our service to one of those Car Wash places that do everything from valet, steam clean and sales as well. He had it done, was charged £15.00 and, of course, after that the car would not start. He called me, said as it was serviced two days ago and the fault was our responsibility. He strangely forgot to mention the steam clean. The RAC recovered the car to us. The electronics were a mess, the cleaners had literally sprayed over the water with WD40, tried and flattened the battery, and the keys were locked in the car as well. We gained entry to the car, checked the electrics, repaired the waterlogged ECU computer, and attended to the wiring problems. A very large invoice indeed. All that's shiny isn't gold, my friend, it just might cost you some.

ALL THAT GLITTERS

Speaking of gold, one of our client's sons turned twenty-one and decided that he wanted to paint his engine gold, so he cleaned the engine, got gold spray paint and did the job. The car was recovered to us with a very sad burnt-out wiring loom and a terrible smell. This plonker had sprayed the engine, manifold, everything, with low temperature paint. As it got hot on the manifold the solvents in the paint caught alight. It took about ten hours' work to rectify the problems, which included engine removal and wiring loom repair. Finally we sprayed the engine properly with UHT (ultra high temperature) paint.

SAD SUSPENSION

A Mercedes S500 came in, with the client claiming that the suspension had gone because it was leaning to the right, and it looked very odd. But it was very simple to rectify, because he'd been sold the wrong size tyres. I was not sure who was the bigger prat: the salesman for selling the wrong size of tyre, or the owner for not noticing!

PUNCH-UP OVER AN INVOICE

It isn't just mechanics who get their pride hurt from time to time. We quote for some work, and the client agrees. The car gets serviced, brakes replaced, all done as promised. When this person came to collect his car, he was arguing with our receptionist that the amount on the invoice was not the same as the quote. I came into the reception as I heard raised voices. I went over the bill, checked the time log and it was correct. He insisted the quote was £75 cheaper than the invoice, and when I showed him the time of the call etc., he said, 'Are you calling me a liar?' Well I wasn't, just that he must be mistaken. With that this person goes to grab my collar, but I'm no pushover so on his attempt I pushed him away and said don't be silly, don't touch. He then tried to get behind the counter so I had to restrain him. The police were called and he wanted to have me arrested for assault, but there were plenty of witnesses who said he was in the wrong. Said fool goes off in a taxi, returns four days later to pay the bill and I put four days storage on as well. No apology and he drove out without making eye contact with anyone.

THE INTERNET IDIOT

One of the funniest things I've seen recently was an old Mitsubishi Pajero diesel, where the client decided to do a self-fitted conversion, courtesy of info he'd got off the internet. He fitted a heater to the fuel filter lines and decided to run the car on old cooking oil. Now, what was not explained is that cooking oil contains particulates of all it has been cooked in, and requires careful filtering with a 350 mesh sub-micronic filter. He simply put in the old cooking oil and hoped it would go. Well it did for about a mile, then stopped and wouldn't re-start. When it was towed in the fuel filter was blocked with crud, it was really smelly and messy. He got the old oil from a chip shop, so you can imagine. We cleaned out the whole diesel system, fitted a pre-heater, in-line heater and filter to stop the oil waxing on cold mornings. We then got 25 litres of cooking oil and all ran sweetly. The client now filters his fuel properly, the only problem is that when you smell the exhaust it makes you hungry. You wonder if he's driving or cooking!

FUEL FIASCO

Cooking oil is one thing, but as you can imagine we often see diesel cars filled with petrol. What you wouldn't believe is the number of times the culprit swears blind they'd never make such a mistake. One guy not so long ago had a Land Rover Discovery Diesel, a nice car, but would it start? No way. Cough, splutter, clouds of smoke. We scan coded it and nothing. So we set about all the normal checks until a fuel pressure test was done. Lo and behold, it was petrol. I called the client who was quite arrogant, offended that I would ever suggest he would make such a mistake and insisted he was coming down to 'sort me out'. He arrived, not happy, and a few words were exchanged between us, until I asked for the fuel receipt, which he did not have. I asked what filling station he went to, and it turned out to be the local Shell garage, whose proprietors happen to be friends of mine. A quick call and they were able to confirm he filled with petrol. So after a flush out, replace filters, tank flush, injector seals etc., and £855 plus £50 of diesel, he apologised (with a red face) and paid. He had driven it about three miles from the filling station and ignored flashing dash warning lights and clouds of smoke from the exhaust. They are embarrassing mistakes, sure, but don't try to deny it. Petrol in a diesel engine is not the sort of thing you can blag.

THE FLASH-FLOOD FOOL

We've all seen James Bond's amphibious car, I reckon, and quite fancied one, but some have been more eager to get their cars into water than they should have been. A car was towed in after the driver/diver decided he could drive his Ford Focus through a brook that had formed in bad weather where other cars had got stuck. This wally drove at full throttle at the puddle, resulting in the air-intake drawing vast amounts of water into the engine. You cannot compress water so it blew the engine. On stripping it I couldn't believe how much damage had been done to the pistons and valves, and a full rebuild was carried out. The car wasn't the only thing that needed drying out – his wallet did too!

THE WRONG TRANSMISSION FLUID

Some of my exploits have seen me ending up in court, though not for the reasons you might think. I had fitted a rebuilt transmission to a Range Rover Sport, and thirteen months later it was brought back because it was slipping in 'drive' gear. We always insist that the car be brought back after 1,000 miles for another transmission oil change to comply with the warranty. The car had covered 19,553 miles since the rebuild and NOT brought back for re-checking. We checked the transmission and it was slipping in drive. We changed the transmission oil and my mechanic asked me to look at the oil he was draining. This is not the correct oil he states, it's very dark and smells acrid. The correct oil is stated by the manufacturer, which is the only one we use. So I called the client and asked him why he did not return after the 1,000 miles to check the transmission. He was very evasive and said he didn't think this was necessary, and his garage had told him this was not required. I asked him if he'd had a transmission service done recently, and his reply was, yes, in France, when it was serviced. I told him that incorrect oil had been used, and also it was not under warranty, since he did not return the car for the oil change. He put down the phone and the next day I received a solicitor's letter demanding I do the repairs under warranty. I wrote back explaining that using the incorrect oil and also not returning for the 1,000-mile re-check made the warranty void. It went to court, I represented myself before the judge, and the case was settled in my favour with my costs to be paid. Reluctantly, he paid to have the transmission rebuilt again and in turn tried to sue the French garage for using the wrong oil which had effectively destroyed the transmission.

TYRE TRAGEDY

All of these cock-ups could've resulted in a serious accident, though thankfully none of them did. But we lost a client a few years ago, not through old age or illness because he was only twenty four. He was the son of one of my long-standing clients who decided to do some repairs himself and save daddy paying for it. Apparently he fitted a CD player and also replaced the wheels and tyres with bigger units (second-hand ones, bought over the internet). The car was a VW polo, ten years old, and on a fast motorway test drive the tyres separated from the rims, causing the car to turn over, resulting in a fatality. The wheels and tyres were of a cheap nondescript make which literally disintegrated. I was really saddened to hear this as on the few times I had met him, he seemed to be a really nice lad. Remember not all tyres are the same – you get what you pay for, and spending a bit more money can sometimes save you your life in the long run.

Who knew that the MOT Inspectorate could also be numpties sometimes? When I was an MOT tester I was real strict – I even failed my wife's car over a torn CV boot!

One day I failed a Ford Cortina for excessive movement of

the driver's door hinge. The door was literally falling off. The owner made a complaint to the MOT Inspectorate, who came to the garage with the client to check my work. The MOT guy tells me I was being too strict and not following guidelines, which I don't take too kindly to as you can imagine, so I tell him to try it for himself. The guy opens the door from the outside, the hinges break away and the door lands on his foot. He gets a broken toe, at which point I mention that steel-cap work boots are required in the workshop. Red-faced he upholds the MOT fail. Well, he could hardly let it back out onto the road with a door missing!

As well as MOT inspectors, I've had my run-ins with the police over the years. I remember one occasion when I felt I was in one of those American movies, you know, when the copper looks over the car looking for any excuse to pull someone in, so smashes their lights with their truncheon and says, 'Oi, no headlights.'

I rebuilt, with my team, a beautiful E-Type Jag, 1966, and when it was finished I took it for a road test. I was stopped by the police, who said they were carrying out roadside checks and added, 'You were going very slow, so you looked suspicious.' They checked the handbrake, tyres, lights etc., and then told me the car was not roadworthy. I said I'd just rebuilt the damn thing, so he said put the handbrake on, and then pushed the car and it moved. I explained that it was a fly-off handbrake, so I put it on and said, 'Now push it.' He couldn't.

Embarrassed, he said my main beam lights weren't working, so I flicked the toggle switch on the dash and the main headlamps worked. Now he was getting desperate. In vain he said the wheel bearings were loose. He put his hand on the tyre and pulled it towards him to prove it, so I pointed out that he was just flexing the tyre, not the bearings. I jacked up the car in the road and showed him. Unable to find anything else he took off in a huff, so I took his number and made a complaint. I received a letter of apology: the letter of the law, you might say.

Here's a real big cock-up – literally! A Jensen Interceptor 1972, four-wheel drive with the Ferguson FF unit. It was a lovely car but there was a burning smell when you were driving. Our mechanic checked all the usual things and could find nothing. So he went on a test drive. One hour later I get a call from him. He is in hospital with burns to his crotch and the car is a total loss. WTF?!?! Turned out that apparently some time previously there was a wiring problem which the client did not make us aware of. The previous garage re-wired the loom under the driver's seat, the wiring chafed, wore away some of its insulation and caused short circuits and sparks and caught the seat alight, leaving our mechanic with a burnt cock and legs before the fire brigade totalled the car.

One of our clients could only be contacted by text on her phone as she had meetings all day. We cannot do any work without approval from the client, so our new nineteen-year-old Polish receptionist sent a quote by text. We know the dangers of predictive text – sometimes you need to check what you write – so what she sent and what she meant were two different things. She copied the mechanic's list as he wrote it and thought nothing about it.

TEXT to client:
Dear Mrs *****
Enclosed is the requirements for your Toyota Prius
Reg, **** ***
Service and lubrication £250 plus VAT
NS front wheel bearing fucked, £88.30

OS front wheel bearing fucked, £88.30

Air pollen filter blocked with crap, £22.66

Bodywork estimate required as both front and NSR panels are dented (must be a good driver!)

All prices include VAT and labour please text your approval asap from ******

Oh my! There was a lot of embarrassment and plenty of apologies by us and the receptionist, who was in her first week in the job. We decided not to fire her as we guessed things could only get better from there on!

Some issues aren't anyone's fault. It just goes to show that no matter how expensive your car is, no car is immune to the hazards of the road, nor can you account for every eventuality:

- A Rolls-Royce Silver Shadow comes in creating a terrible smell when driving – acrid it was, it made your eyes water. I thought something was seriously wrong with this baby. Checked in the engine bay and interior, fault located to a large polythene bag stuck to the exhaust, so when it's hot the plastic produces the smell. Chiselled off the bag and no more smell.

- My head-scratching wasn't the only scratching going on here. Bits of paper and fluff were appearing constantly in this Ford Mondeo. I could not trace any problem, so we removed the dashboard, only to find a mouse living behind the dash who was having dinner on the air duct hoses. The mouse was removed and released to fend for itself on the mean streets of Golders Green, the dash refitted with new air vents and no more fluff.

- When a client says he is using vast amounts of fuel, there are certain things you can check. This Jaguar XF, according to the fuel record, was going through about four miles per gallon. I told the client to record his fill-ups and mileage after we did a scan code that showed no faults. The next week was the same recording, about three to four mpg, but the mileage was showing hundreds more than when he left it. The gauge was checked and it was OK, there still no faults apparent. Eventually the guy worked out that his son was secretly using the car during the day whilst he was at work, so the mystery was solved. He kept both sets of keys on him at all times in future!

Occasionally you get a client who tries it on, but when a person comes in with an attitude, I get hot under the collar. Once, this guy comes into reception all suited and booted, spotless fingernails, and a face like a smacked arse.

'I wish to make a complaint,' he says. 'You serviced my car five weeks ago, at vast expense, and now it won't start.'

'OK,' I say, 'what's your name please?'

He says, 'You should know me, after all the money I've spent here over the years.'

Now, I am very good with faces, but his face don't ring my bell. Again I ask for his name, and again he is getting shirty, saying he's a regular, but it's Mr ******.

So I checked the name on our database, and LO AND BEHOLD, the last invoice is 2011 and it's now 2014! Don't try and pull the wool over my eyes, sunshine! He left the garage in a huff, never to be seen again. I hate attitudes. You give 'em to me, and you get 'em right back.

Talking of attitudes, if there's one species that's got 'em it's wheel clampers. Big, fat grumpy blokes with nothing better to do than ruin other people's days. My daughter Nadine, when she had her young baby, stopped in a parking bay behind some shops while she ran in with baby to grab some things. She returned after ten minutes, and she was clamped. She saw no signs, finally locating one over twenty feet up a wall. The shop owner knew nothing about it. The clamper, the usual rather large guy with an attitude, told my daughter it's £250 to remove the clamp, and after thirty minutes her car would be towed away, so there would be an extra £150 for towing plus storage.

She called the police who attended and said they could do nothing, as one of the flats above the shop had asked the clamp company to do this to stop people parking. She called me, and I sent out one of my recovery trucks. My driver Dennis is a very big lad – six-foot six inches, and very well built. He calmly put wheel skates on Nadine's car, attached a hydraulic winch and lifted the car onto the flat bed.

The clamper looked on in amazement. Dennis handed the man a card and said, 'If you want your clamp back, it's £250 plus storage.' Then he drove away. The clamp company came to the garage three days later, apologised and paid the fee. The clamp company are no longer on this patch.

Finally, there's good old Leepu. Where would I be without him? I'd probably have some fucking hair for a start! I love the guy, but my God he can be an idiot sometimes. You can guarantee if there's a tough job that needs doing, Leepu will find a way to make it that little bit harder for mechanics. Here are some moments when I could've quite happily put my hands round his fat neck and strangled him:

- Leepu has always loved cutting chassis and panels with little thought for the result, except for the design. All he cares about is what a car looks like and in his determination to get *his* job done, he forgets about everyone else's. The car we built for Jools Holland, the famous blues pianist, was chopped at the roof, as I mentioned before. The result was that with no strengthening, the body bent and resulted in over fifty hours of extra work for myself and the team to straighten and strengthen it again. Meanwhile, Leepu went home for dinner!

- Another time we were modifying a Jaguar XJS. I was working underneath the car and 'Chopper', as I called Leepu, decides to cut the roof off. The idiot cuts straight through the wiring loom, all twenty-three wires of it! He looks at me and I look at him, he can see I'm really angry, and all he can say is 'Whoops!' Result? Me chasing him from the garage with a spanner, and he never returned for two days! But we kissed and made up eventually, as we always did.

CHAPTER SIXTEEN

BERNIE'S TIPS AND TRICKS

There are lots of ways you can help yourself save money at the garage and get yourself out of a pickle when you most need it, and you don't need any special skills or professional equipment to put them into practice.

AIR CONDITIONING

Over time the moisture in your air con will naturally create a build-up of algae in the air vents. Some garages charge anything from £50 to £250 to service air conditioning systems and rid you of the smell that the algae creates. My method costs pennies and will neutralise and get rid of the bacteria. Get a pipette from any good chemist and apply 'Ylang Ylang' or lavender essential oils (or an essential oil of your choice). Add a few drops into all the air vents, about ten drops per vent, and allow to settle for two hours before using the air conditioning again. When you do you will get a much more pleasant smell and no more algae bacteria. This tip will also help people with travel sickness.

HEATER NOT KEEPING YOU WARM?

If you find your heater isn't keeping you warm or creating a smell like burning paper, it probably means the pollen filter needs a clean. Simply take it out and clean or replace the thing. This will save you hundreds of pounds on a new heating system.

BROKEN RADIATOR HOSE

It's one of those things on a car that can go at any moment, but there is a simple fix if you ever find yourself on the road and discover water leaking from your radiator hose. Simply carry duct tape in the boot of your car. Apply this around the hose as tightly as you can, with four or five turns to seal the hole completely. It's always a good idea to carry a bottle of water with you too, but if you don't have one or access to one, then you can fill up the radiator with the contents of the screenwash bottle, just as a temporary measure to get you to the nearest garage. If it's a slow leak, then these measures will (when driven slowly) get you to the nearest garage to be able to effect a replacement hose. In the trade these are called 'running repairs'.

WARNING: ALWAYS BE CAREFUL WITH THE HOT WATER IN THE RADIATOR. SCALDING WATER CAN GUSH OUT AT PRESSURE IF YOU REMOVE THE CAP.

A NEW PAINT JOB

If you've got dull paintwork then your only options are a re-spray or a full paint valet, both of which can cost you thousands of pounds. But with my way you can get great results for barely ten pounds. Purchase some good liquid car polish and exfoliating face cream. Mix together ten parts polish to three parts cream. Polish into each panel and allow to dry, then buff to a superb shine. Dead paint gone, shine back and hundreds of pounds saved!

CHIPPED PAINTWORK

Small lacquer chips on your car paintwork can be invisibly mended using clear nail varnish. Clean the area with a clean cloth (NOT polish) and, using the brush supplied, gently paint over the lacquer chips. As if by magic they disappear.

CHROME WORK

If your chrome work is dull then the technique is slightly different. Use a dry bit of wire wool on the rust and rub all around the chrome. Amazingly this will take away all the pitting and your chrome will get a new-look shine. Then simply finish off with a good car polish to get a gleam and to give it some protection.

PLASTIC TRIMS

If you have plastic trims rather than chrome work and it has dulled, simply use some baby wipes to clean and shine it. You can also use baby wipes on the dashboard and plastic door panels to make them look like new again.

WINDOWS

If it's grubby or streaked windows that are letting the look of your car down, simply spray a little water onto the glass and buff off with some crumpled newspaper. You'll be surprised with the results!

DEAD BATTERY?

Before you replace a dead car battery, you may want to check the water levels. What?! Yes, batteries have distilled water in them to make the chemical and electrical reaction to charge a battery. This distilled water running low can sometimes be the cause of a flat battery, but purchasing distilled water isn't cheap. As an alternative you can create your own and there are several ways of doing this. The easiest is perhaps to quarter-fill a stainless steel saucepan with tap water. Place a glass bowl on top so that it is floating in the water. Place the lid to your stainless steel saucepan on top, but upside down so that you can place some ice on top of it. Put the saucepan and the contents onto the hob to boil, then add the ice into the lid. This will make the lid cold, so that when the steam from the boiling water below hits it condensation is created, which will then drip back into your glass bowl, but minus the impurities.

This de-ionized water can then be used to top up the plates inside your battery. Put it on a slow charge using a battery charger, and in a few hours your battery will have a new lease of life!
WARNING: BE CAREFUL WITH BOILING WATER.
ALLOW TO COOL COMPLETELY BEFORE USING
YOUR DISTILLED WATER.

CIGARETTE LIGHTER

If you find your cigarette lighter not working it might be the element. If this corrodes and doesn't make contact then it won't work, so remove the element and clean it with an old toothbrush. This could save you up to £100 for a replacement cigarette lighter.

OIL LEAKS

If oil leaks or spills are making a mess of your expensive driveway you just need to sprinkle some cat litter over the offending splodge, rub it in and leave overnight. When you sweep up the next morning the oil will be gone and your drive should be clean.

SICK STAINS

Why do other people's children wait until they are in your car before being sick? If they do have a mishap on your car carpet, once you've removed the excess, rub in some lemon juice and this will help to get rid of the lingering smell.

SQUEAKY BRAKES

There are few things more annoying about motoring than a car with squeaky brakes. If you do get them, even on a new car, it's worth trying this little trick. Find a straight, empty road, making sure the way in front of you and behind you is COMPLETELY CLEAR. Drive at 30 mph with one foot on the accelerator and lightly touch the brake at the same time. This will skim the top surface of the brake pads and hopefully stop the squeak.

WATER IN YOUR HEADLAMPS

If you find that water in your headlamps is causing a distorted beam, first find the hole that is causing the leak and seal it by lightly applying bath sealant. Then dry out the water with a hairdryer, holding it about two inches away from the glass until the water disappears.

TYRES

If you find a tyre keeps losing air, spray around the valve and wheel hub with diluted washing up fluid. The bubbles will show you where the leak is. It may just need a simple repair, which can save you buying a whole new tyre.

PINKING

Do you have an older car that is pinking (misfiring) on acceleration and you can't find high-octane fuel? Add a couple of simple non-coloured mothballs to the tank. They will dissolve in seconds and increase the octane rating by as much as 5 per cent. Result? More performance and no pinking, and less money needed to pay for high-octane fuel.

ELECTRICS

If you find fault codes appearing on the dashboard of your vehicle but no obvious problem, try this tip before you spend hundreds of pounds on diagnostics. Disconnect the earth wire on the battery, leave it for twenty seconds then re-connect it (without the keys in the ignition). Sometimes electronics get a glitch in the system and this can re-boot it, often curing the fault.

JAMMED DOORS

Jammed and seized door locks in the winter? Heat the tip of the key with a cigarette lighter for a few seconds, then push slowly into the lock: Magic! Then add a suitable releasing fluid to the lock such as WD40. The result? No more seizure.

WHEEL NUTS

If you find your wheel nuts are very tight and you cannot undo them with the supplied nut wrench, try tightening them up further before trying again. Or safely stand on the end of the wheel wrench whilst holding on to the roof channel.

SEIZED BOLTS

Have you got seized-up bolts, and no penetrating fluid to cut through the corrosion? Boil some vinegar, apply it to the bolts or nuts, leave it for a few seconds and, as if by magic, they will undo.

DOOR RUBBERS

It can be very annoying to have wind whistling through the door rubbers when you are driving at high speeds. Clean the door aperture rubbers with soapy water and thoroughly dry them. Apply petroleum jelly to the whole perimeter of the door rubbers. This will make them supple and will seal, resulting in no more wind noise. This works great with convertible roof rubbers as well.

STATIC SHOCK

If you are continually getting static shocks, hold onto the body of your car when getting out. The static charge will be grounded through the tyres, and you'll get no more shocks.

CIGARETTE BURNS

Small cigarette burns in carpets and seals can be mended with ash – yes, I said ash. Apply a small amount of superglue to the burn, powder some ash onto this and let it dry. Apply a suitable dab of paint to this when it's dry for colour matching.

FUEL CONSUMPTION

Using more fuel than you should be? Check the tyre pressures are correct. Nothing wastes more fuel than incorrect pressures or over/un-inflated tyres.

DOG DAYS

We've all had trouble with our motors in the winter, but sometimes hot weather can cause problems too. If you find your car won't start on a particularly hot day then remember that fuel vapour locks are common. Place a bag of ice or frozen peas or something similar on the inlet manifold, leave it for ten minutes, and the vapour lock will go, and the car will start.

BERNIE'S ALL-TIME TOP 10 CARS

Over all these years I have had the privilege to work on some of the finest machines created by man. Cars at their very best are things of beauty, both the bodywork and the hidden craftsmanship and design under the bonnet. The great cars I've had the pleasure to drive and handle are too numerous to mention, but here are ten cars that I think, in their day and in their own quirky way, can stand alongside any car ever built.

1. FORD ZODIAC CONVERTIBLE CIRCA 1959
This is the epitome of the finest Ford design. A no-nonsense 6 cylinder, 2.5 litre engine that's so simple even Stevie Wonder could work on this car, with its 3-speed manual transmission, hydraulic power roof, four full seats, and a bench seat for getting the girls close for a snog and more. Beautiful lines, open top, a real crowd puller. Had my first leg over in one of these cars, so you could say I know it intimately!

2. JAGUAR MK 2 1964, 3.8 AUTOMATIC
A real 'hoods' car, all the real gangsters had one of these. It has stunning looks, a powerful 3.8 twin cam engine, luxury leather interior and wire wheels. If you drove one of these you were the mutt's nuts, a real geezer.

3. TRIUMPH VITESSE CONVERTIBLE
2.5 straight 6 cylinder engine, twin carburettor, real power

and good looks. A summer car like no other, with its four seats, easy front-lift bonnet, and you could change the clutch from inside the car by removing the gear shroud. I had one and loved it.

4. MGB GT

Oh, those memories! Great British engineering, simple 1800cc engine, 4 speed with Laycock overdrive. A proper GT grand tourer car – mine was black with grey stripe interior. I changed the body to half black, half silver, and with the Rostyle wheels it was a stunner.

5. JAGUAR E-TYPE

What can I say about this car that hasn't already been said? It's an absolute icon. A true sports car, twin cam 6 cylinder engine (not the V12... yuk!), superb lines, a real bird puller, a noted first true 160 mph sports car. Looks like a penis on wheels, the greatest shape of any car ever designed.

6. JENSEN INTERCEPTOR

Big V8 engine, torque flight auto transmission, Pininfarina styling and made in the UK. With its leather interior it was true luxury. The classiest car I ever owned.

7. MERCEDES C270 CDI DIESEL

Real V6 diesel power, ultra luxury two-door coupé, stunning performance, real leather interior. True German engineering at its best.

8. JAGUAR 4.2 COUPÉ

White with black leather interior, auto transmission, two-door with four electric windows, who needs air conditioning with this luxury classic car? I had one for three years, but the missus always used it, and I never got a look-in!

9. RANGE ROVER P38 MK 2
Year 2000 with Thor V8 4-litre petrol engine. A fantastic car, a bit juicy, but so many luxury refinements: leather, heated seats, cruise, power, 4-wheel drive, once you iron all the problems out it is the best. My daily drive.

10. CORVETTE STINGRAY CIRCA 1968
The true American Muscle Car, it's like me: built for comfort not speed. Superb lines, stunning looks, handles like a boat on corners, but a real crowd puller. I've had some fun with these over the years!

ACKNOWLEDGEMENTS

A man is not a man without his family and friends.

To my loving devoted wife, Lisa, who has stood by me through the good and bad times, fired me up, made me what I am today, I truly love you.

To my parents-in-law, Jo and Des, love and respect to you both.

To my daughters, Nadine and Lisa Marie, my sons-in-law, Jamie and Gareth, I am so very proud of you, truly with love from my heart, God bless you all.

To my grandchildren, Chloe, Daisy and Nia, I am the proudest Papa Grandpa in the world.

To my best friends any man could ever want: Phil and Sharon, Gerry and Shirley Rozalar, Christopher and Cil, you have been my rock; Wayne and Sue Morris, and my goddaughter Erin, Lyndsay and Colin Paice.

Mike and Ross Schofield and family, Colin Lincoln, Simon and Tommy (mechanics) Wayne (Baldy) – the best paintworker in the UK, George (the central European), Jake, Brett Curran and Big Clinton, Izzy, Duncan and Lindsay, Scottydog Manning, Beth, Neal, Harry and all my friends at the Samaritans, Ismet (Arches Café), Sean Hart, Andy (Oompa Loompa) Summers, Lee Essex, Phil (get the fuck outta here) Butler, Dr Rai (for keeping me healthy) and Dr Antony Karial (my dentist at the Gingerbread House).

Jimmy Nunn (still friends after 55 years), Dave (Fingers) Smith, Lenny Bonham, Eddy Pring (hard bastard), Denny Gilson (the fixer), Giancarlo Ceri (Miami), Dilip Shah, Stevie Wilson (Pork Pie), Terry Slinger (I'll punch yer lights out yer wanker).

To my great friends in TV who helped my dreams come true, Leepu, Steve Mizalas and all the team on *Chop Shop*, and Mario, Steve Scott and the *Classic Car Rescue* team.

To Wally Strong and Ted Watson, the master mechanics who guided me, taught me, helped me read and write, and made a mechanic out of me, God rest your souls.

To my friends and family: thank you for everything you do for me, bless you, and Tommy my Chihuahua, thank you for showing people what a big softie I am really.

I would also like to sincerely thank my agent Andrew Wilson, of Cloud9 Management, for his understanding, professionalism, and for writing this book from my assorted bits of paper, bad English, swear words, rants, late night calls, and nagging.